PERFORMING IN EXTREME ENVIRONMENTS

Lawrence E. Armstrong, PhD
University of Connecticut

Human Kinetics

Library of Congress Cataloging-in-Publication Data

Armstrong, Lawrence E., 1949-
   Performing in extreme environments / by Lawrence E. Armstrong.
     p.   cm.
   Includes bibliographical references (p.   ) and index.
   ISBN 0-88011-837-7
   1. Sports--Physiological aspects. 2. Extreme environments-
-Physiological effect. 3. Exercise--Physiological aspects.
I. Title.
RC1238.A75 2000
612'.044--dc21   99-038795
                                              CIP

ISBN-10: 0-88011-837-7
ISBN-13: 978-0-88011-837-8

Acquisitions Editor: Michael S. Bahrke, PhD; **Developmental Editor:** Rebecca Crist; **Assistant Editor:** Chris Enstrom; **Copyeditor:** John Mulvihill; **Proofreader:** Sue Fetters; **Permission Manager:** Terri Hamer; **Graphic Designer:** Fred Starbird; **Graphic Artist:** Denise Lowry; **Photo Editor:** Clark Brooks; **Cover Designer:** Jack W. Davis; **Photographer (cover):** PHOTOWORKS/Gay Wiseman; **Illustrators:** Mic Greenberg (Mac) and Kristin Mount (line and medical drawings); **Printer:** Versa Press

Human Kinetics books are available at special discounts for bulk purchase. Special editions or book excerpts can also be created to specification. For details, contact the Special Sales Manager at Human Kinetics.

Printed in the United States of America   10 9 8 7 6

Human Kinetics
Web site: www.HumanKinetics.com

United States: Human Kinetics
P.O. Box 5076
Champaign, IL 61825-5076
800-747-4457
e-mail: humank@hkusa.com

Canada: Human Kinetics
475 Devonshire Road Unit 100
Windsor, ON N8Y 2L5
800-465-7301 (in Canada only)
e-mail: orders@hkcanada.com

Europe: Human Kinetics
107 Bradford Road
Stanningley
Leeds LS28 6AT, United Kingdom
+44 (0) 113 255 5665
e-mail: hk@hkeurope.com

Australia: Human Kinetics
57A Price Avenue
Lower Mitcham, South Australia 5062
08 8372 0999
e-mail: liaw@hkaustralia.com

New Zealand: Human Kinetics
Division of Sports Distributors NZ
Ltd.
P.O. Box 300 226 Albany
North Shore City
Auckland
0064 9 448 1207
e-mail: info@humankinetics.co.nz

DEDICATION
AND
ACKNOWLEDGMENTS

This book is dedicated to family members who influenced my childhood and enriched my adult life: Edward, Louise, Lynn, Lee, Lori, Joyce, April, Jim, and grandchild Jamie.

To Carl and Jaci, my close friends and colleagues, and to the numerous graduate students at the University of Connecticut who have enjoyed research, writing, and learning as much as I do.

To those who encouraged and assisted me during the production of this book: Don, Deborah, Elaine, Heather, Helen, Jeff, Joyce, Judy, Pamela, and Melinda.

Dr. Leo Senay and Dr. Stavros Kavouras read this manuscript and offered numerous helpful editorial suggestions.

After recognizing the numerous marvelous ways that human systems respond to the earth's environments, I am compelled to agree with the following verse: "The earth is the Lord's and everything in it, the world and all who live in it; for He founded it upon the seas and established it upon the waters." (Psalm 24:1-2)

CONTENTS

PREFACE

Earth, third planet from the sun, contains an awesome array of surface environments, ranging from Antarctica to the Sahara Desert and from Mount Everest to the Mariana Trench. Although each locale presents danger to human life, we attempt to explore, work in, or inhabit these extreme environments. Physical training and athletic competition are no different, in that some individuals exercise in extremes of temperature, humidity, solar radiation, wind speed, altitude, and water pressure.

In response to exercise in severe environments, the human body may be able to maintain its temperature, fluids, and chemicals in balance until the exercise ends. If not, fatigue or illness can result. It is also possible that the body will adapt to repeated exposures by a decrease in heart rate, core body temperature, or fatigue. But one thing is certain: each of the earth's environments affects the body in a unique way.

Even ordinary seasonal changes can be problematic. During the past five years, Connecticut residents have experienced air temperatures ranging from 38°C to –25°C (101°F to –13°F), humidities from 18% to 100%, and hurricane force winds on several occasions. We also should not forget the plaque at the peak of Mount Washington, New Hampshire, that memorializes the day on which one of the fastest wind speeds on earth was recorded—231 mph in 1934! Despite this, New England usually is not considered to have a stressful climate. Without an understanding of the dangers that exist in our environment and the required countermeasures, athletes (a) will not optimize training or performance, and (b) run the risk of incurring illness or injury that could impede or terminate their athletic pursuits.

Therefore, this book is written to provide advice to runners, cyclists, climbers, divers, recreational enthusiasts, soldiers, and laborers, as well as high school, college, and professional athletic teams. It is necessary because so many people understand *something* about environmental physiology, but few know *enough* or take the precautions that are necessary. Further, because training and competition are usually based on trial and error or tradition, this book has been written to provide objective information from the fields of environmental physiology, exercise physiology, and environmental medicine.

It represents a distillation of decades of research (from other laboratories and our own) and thousands of interactions that I have had with athletes, coaches, and soldiers. However, care has been taken to provide practical advice in words that can be understood by those who are unfamiliar with physiology and medicine.

The organization of this book allows the reader to examine all of the earth's stressful physical factors (i.e., heat, humidity, air pollution, cold, windchill, high altitude, day length, air ions, and underwater pressure). Chapters 2 through 8 provide details about the body organs that are affected, ways that physical factors alter exercise performance, medical disorders, and tips to avoid performance decrements and illness. When faced with exercise or labor in hostile environments, the reader will learn that adjustments can be made in behavior, training patterns, eating habits, fluids, clothing, medications, or the air mixture breathed.

If the reader enjoys optimal performance, remains healthy during and after exercise, and understands physiological processes as they occur, the purpose of this book will be fulfilled.

CREDITS

Fig 1.3   Adapted, by permission, from R.M. Berne and M.N. Levy, 1996, *Principles of physiology* (St. Louis: Mosby), 694.

Table 1.1   Adapted, by permission, from N.H. Spector, 1990, "Basic mechanisms and pathways of neuroimmuno-modulation: Triggers," *International Journal of Neuroscience* 51: 335-337.

Table 1.3   Adapted, by permission, from H. Weiner, 1992, Definition and classification of stressful experience. In *Perturbing the organism. The biology of stressful experience* (Chicago: University of Chicago Press), 37-55. Copyright © 1992 by The University of Chicago.

Table 1.4   Adapted, by permission, from W.D. McArdle, F.I. Katch, and V.L. Katch, 1996, *Exercise physiology: Energy, nutrition, and human performance* (Baltimore: Williams & Wilkins), 362-364.

Fig 2.2   Adapted, by permission, from L.E. Armstrong, 1991, "Environmental considerations: Body temperature regulation," *NSCA Journal* 13 (3): 66-67.

Fig 2.3   Adapted, by permission, from L.E. Armstrong, 1992, *Keeping your cool in Barcelona: A detailed report* (pamphlet) (Colorado Springs, CO: Sport Sciences Division, USOC), 6.

Fig 2.4   Adapted, by permission, from L.E. Armstrong, 1992, *Keeping your cool in Barcelona: A detailed report* (pamphlet) (Colorado Springs, CO: Sport Sciences Division, USOC), 9.

Fig 2.8   Reprinted, by permission, from L.E. Armstrong, 1998, "Urinary indices during dehydration, exercise, and rehydration," *International Journal of Sport Nutrition* 8 (4): 349.

Fig 2.9   Reprinted, by permission, from L.E. Armstrong et al., 1996, "Heat and cold illnesses during distance running," *Medicine & Science in Sport & Exercise* 28 (12): ii.

Fig 2.10   Reprinted from *Adaptation Biology and Medicine,* Volume 2. Editors: K.B. Pandolf, N. Takeda and P.K. Singal. Copyright © Narosa Publishing House, New Delhi, India, 1999.

Fig 2.11   Reprinted from *Adaptation Biology and Medicine,* Volume 2. Editors: K.B. Pandolf, N. Takeda and P.K. Singal. Copyright © Narosa Publishing House, New Delhi, India, 1999.

Fig 2.12   Reprinted, by permission, from W.B. Bean and L.W. Eichna, 1943, "Performance in relation to environmental temperature," *Federation Proceedings* 2: 144-158.

Table 2.1   Adapted, by permission, from L.E. Armstrong, 1992, *Keeping your cool in Barcelona: A detailed report* (pamphlet). (Colorado Springs, CO: Sport Sciences Division, USOC), 12.

Table 2.2   Reprinted, by permission, from L.E. Armstrong and J.E. Dziados, 1986, Effects of heat exposure on the exercising adult. In *Sports physical therapy,* edited by D.B. Bernhardt (New York: Churchill Livingstone), 197-214.

Table 2.8   Adapted, by permission, from W.F. Mellor, 1972, *Casualties and medical statistics*, 20-51. Crown copyright is reproduced with the permission of the Controller of Her Majesty's Stationery Office, London.

Fig 3.1   Adapted, by permission, from G.E. Folk, 1966, *Introduction to environmental physiology: Environmental extremes and mammalian survival* (Philadelphia: Lea & Febiger), 97.

Fig 3.2   Adapted, by permission, from D. Milesko-Pytel, 1983, "Helping the frostbitten patient," *Patient Care* 17: 90-115. Copyright © Medical Economics.

Fig 3.4   From B.J. Noble, *Physiology of exercise and sport*, 1986, St. Louis, Mosby.

Fig 3.5   Adapted, by permission, from R.M. Smith and J.M. Hanna, 1975, "Skinfolds and resting heat loss in cold air and water: Temperature equivalence," *Journal of Applied Physiology* 39: 93-102.

Fig 3.7   Adapted, by permission, from E.T. Poehlman et al., 1990, "The impact of physical activity and cold exposure on food intake and energy expenditure in man, "*Journal of Wilderness Medicine* 1: 265-278.

Fig 3.8   Adapted, by permission, from S.D.R. Galloway and R.J. Maughan, 1997, "Effects of ambient temperature on the capacity to perform prolonged cycle exercise in man," *Medicine and Science in Sports and Exercise* 29: 1240-1249.

Fig 3.9   Adapted, by permission, from S.D.R. Galloway and R.J. Maughan, 1997, "Effects of ambient temperature on the capacity to perform prolonged cycle exercise in man," *Medicine and Science in Sports and Exercise* 29: 1240-1249.

Fig 3.10   Adapted, by permission, from M.M. Toner and W.D. McArdle, 1988, Physiological adjustments of man to cold. In *Human performance physiology and environmental medicine at terrestrial extremes*, edited by K.B. Pandolf, M.N. Sawka, and R.R. Gonzalez (Indianapolis: Benchmark Press), 361-400.

Fig 3.11   Adapted, by permission, from K.B. Pandolf et al., 1987, Influence of body mass, morphology, and gender on thermal responses during immersion in cold water. In *Ninth international symposium on underwater and hyperbaric physiology*, edited by A.A. Bove, A.J. Bachrach, and L.J. Greenbaum, Jr. (Bethesda, MD: Undersea and Hyperbaric Medical Society), 145-152.

Fig 3.14   Adapted, by permission, from J. Foray, 1992, "Mountain frostbite," *International Journal of Sports Medicine* 13: S194.

Fig 3.15 (top panel)   Adapted, by permission, from K.C. Parsons, 1993, *Human thermal environments* (London: Taylor & Francis), 181-198.

Fig 3.15 (bottom panel)   Adapted, by permission, from R.R. Gonzalez, 1988, "Biophysics of heat transfer and clothing considerations. In *Human performance physiology and environmental medicine at terrestrial extremes*, edited by K.B. Pandolf, M.N. Sawka, and R.R. Gonzalez (Indianapolis: Benchmark Press), 60.

Table 3.1   Adapted, by permission, from T.J. Doubt, 1991, "Physiology of exercise in the cold," *Sports Medicine* 11: 367-381.

Table 3.4   Adapted, by permission, from K.C. Parsons, 1993, *Human thermal environments* (London: Taylor & Francis).

Table 3.5   Adapted, by permission, from T. Schimelpfenig and L. Lindsey, 1991, Cold injuries. In *NOLS wilderness first aid* (Mechanicsburg, PA: Stackpole Books), 153-179.

Fig 4.4   Adapted, by permission, from D. Graver, 1993, Adapting to the aquatic environment. In *Scuba Diving* (Champaign, IL: Human Kinetics), 79.

Fig 4.5   Adapted, by permission, from R.H. Strauss, 1984, Medical aspects of scuba and breath-hold diving. In *Sports medicine*, edited by R.H. Strauss (Philadelphia: W.B. Saunders Co.), 362.

Fig 4.6   Adapted, by permission, from R.H. Strauss, 1984, Medical aspects of scuba and breath-hold diving. In *Sports medicine*, edited by R.H. Strauss (Philadelphia: W.B. Saunders Co.), 363.

Fig 4.8   Adapted, by permission, from K.W. Kizer, 1995, Scuba diving and dysbarism. In *Management of wilderness and environmental emergencies*, edited by P. Auerbach (St. Louis: Mosby), 1184.

Fig 4.9   Adapted, by permission, from W.D. McArdle, F.I. Katch, and V.L. Katch, 1996, *Exercise physiology: Energy, nutrition, and human performance* (Baltimore: Williams & Wilkins), 535.

Table 4.3   Adapted, by permission, from J.M. Clark and S.R. Thom, 1997, Toxicity of oxygen, carbon dioxide, and carbon monoxide. In *Diving medicine,* 3rd ed. (Philadelphia: W.B. Saunders Co.), 131-145.

Fig 5.3   Adapted, by permission, from A. Cymerman, 1996, The physiology of high-altitude exposure. In *Nutritional needs in cold and high-altitude environments*, edited by B. Marriott and S.J. Carlson (Washington, DC: National Academy Press), 295-317.

Fig 5.4   Reprinted, by permission, from A.J. Young, 1988, Human acclimatization to high terrestrial altitude. In *Human performance physiology and environmental medicine at terrestrial extremes*, edited by K.B. Pandolf, M.N. Sawka, and R.R. Gonzalez (Indianapolis: Benchmark Press), 514.

Table 5.1   Reprinted, by permission, from A.J. Young, 1988, Human acclimatization to high terrestrial altitude. In *Human performance physiology and environmental medicine at terrestrial extremes*, edited by K.B. Pandolf, M.N. Sawka, and R.R. Gonzalez (Indianapolis: Benchmark Press), 512.

Table 5.2   Reprinted, by permission, from A.J. Young, 1988, Human acclimatization to high terrestrial altitude. In *Human performance physiology and environmental medicine at terrestrial extremes*, edited by K.B. Pandolf, M.N. Sawka, and R.R. Gonzalez (Indianapolis: Benchmark Press), 520.

Excerpt on page 197   Reprinted, by permission, from P.N. Frykman, 1988, "Effects of air pollution on human exercise performance," *JASSR* 2 (4): 66-71.

Fig 6.2   Reprinted, by permission, from P.N. Frykman, 1988, "Effects of air pollution on human exercise performance," *JASSR* 2 (4): 66-71.

Fig 6.3   Reprinted, by permission, from S.M. Horvath, 1981, Impact of air quality in exercise

Adapting to Stressful Environments

chapter 1

---

*Life is an example of the way in which an energy-system in its give
and take with the energy-system around it [i.e., the earth] can
continue to maintain itself for a period as . . . a self-balanced unity.
Perhaps the most striking feature of it is that it acts as though it
"desired" to maintain itself.*

—C.S. Sherrington, *Man on His Nature*

Your body constantly, automatically attempts to maintain essential
nutrients and chemicals at normal levels in order to maintain health
and performance. The purpose of this chapter is to describe the short-
and long-term adaptations that humans make when exposed to severe environments. These adaptations involve many body systems,
but the central nervous system and stress hormones (e.g., norepinephrine, epinephrine, cortisol) are especially vital. The terms defined in this chapter are important to a complete understanding of all
environments presented in chapters 2 through 8.

ENVIRONMENTAL CHALLENGES TO INTERNAL STABILITY

Your body consists of materials and elements that are extremely un-
stable. It takes only a few minutes of oxygen deprivation before your

brain cells are irreparably damaged, or a rise in body temperature of only a few degrees Celsius before your proteins begin to break apart at the molecular level. In comparison to the mountains, oceans, and atmosphere of earth, your body is extremely fragile and tentative. Despite this fact, you exist and are in good health because numerous physiological responses and biochemical reactions constantly work to maintain stability inside your cells.

The 19th-century French scientist Claude Bernard is acknowledged as the first author to recognize that humans can disregard the external environment because they maintain constancy inside cells.[1] Bernard named the internal environment of cells the *milieu intérieur* and observed that, by maintaining internal stability, humans are independent of the external world. This explains why you can live, work, and exercise in a variety of potentially dangerous environments—temperate, tropical, mountainous, or frigid—and also explains the illnesses that may result. In Bernard's words:

> *The organism is … constructed in such a fashion that, on the one hand, there is full communication between the external environment and the milieu intérieur, and on the other, that there are protective functions … holding living materials in reserve and maintaining [temperature, fluids] and other conditions indispensable to vital activity. Sickness and death are only a dislocation or perturbation of [these processes].[2]*

An important refinement of Bernard's work was published in 1929 by Harvard physiologist Walter Cannon. He described the actions of cells responding to perturbing stimuli in terms of dynamic equilibria and variability, rather than absolute intracellular constancy. Cannon coined the term **homeostasis** to indicate similarity with some variability, rather than "sameness." Homeostasis refers to the body's tendency to maintain a steady state despite external changes.

UNDERSTANDING THE STUDY OF STRESS

The experimental study of **stress** was pioneered by Hans Selye, a Canadian physician, in the 1950s. Selye's research is very relevant to the present book because he investigated responses to strenuous exercise, prolonged exposure to heat and cold, and several other damaging stimuli, which he named "stressors."[3] **Stressors** are influ-

Table 1.1
Stressors That Disrupt Human Homeostasis and Cause Physiological
Responses or Adaptations

External environment	Internal environment
Heat	Heat
Cold	Cold
Odor	Sleep
Food	Hunger
Water	Thirst
Hypoxia	Infection
Noise	Ion imbalance
Light	Fear, anxiety
Darkness	Muscle tension
Trauma, injury	Internal clocks
Electric shock	Intense emotions
Physical threat	Autonomic change
Bacteria, viruses	Abstract thoughts

Adapted from Spector 1990.

ences that throw your body out of homeostatic balance (e.g., unpleasant or noxious stimuli). Table 1.1 presents many stressors that affect homeostasis. These may be external or internal in nature.

Adaptive responses, or **adaptations** (physiological changes that minimize bodily **strain**), are the body's attempts to counteract stressors and reestablish homeostasis; the return to nonstress conditions reflects improved bodily function in the involved organ or system. Adaptations may be either short-term (accommodation), intermediate in duration (acclimation, acclimatization), or long-term (genetic adaptation). **Accommodation** refers to an immediate physiological change in the sensitivity of a cell or tissue to change(s) in the external environment. Both **acclimation** and **acclimatization** involve a complex array of adaptive responses, but the former is induced experimentally in an artificial environment whereas the latter is induced by exposure to natural environments. **Genetic adaptation** refers to semipermanent morphological, physiological, or other changes that occur over many generations within one species that favor survival in a particular environment. Table 1.2 compares the types of *acquired* and *inherited* adaptations that allow humans to survive on earth. Examples of these include enhanced sweating after heat acclimatization

Table 1.2
Acquired and Inherited Adaptations That Allow Humans to Maintain Cellular Stability, Good Health, and Optimal Performance

Mode	Site	Affect	Time required to develop	Examples
Acquired during life	Cells	Biochemical reactions	Minutes, hours	Hormone secretion, change in enzyme activity, change in excitability, hormone receptor up or down regulation
	Tissue, organs	Structure (e.g., membranes, cytoplasm)	Days	Hypertrophy, hyperplasia
	Organ systems, body	Function	Days, weeks	Heat or high-altitude acclimatization
Inherited at birth	Population	Traits (i.e., genotype)	Years, generations	Changes in the characteristics of a group or species, natural selection

From *Handbook of Physiology: Section 4: Environmental Physiology Two Volume Set*, edited by Melvin J. Fregly and Clark Blatteis. Copyright © 1996 by American Physiological Society. Used by permission of Oxford University Press.

(i.e., acquired) and the great body fat insulation found in Inuit peoples (i.e., inherited).

PRINCIPLES OF EXPOSURE TO STRESSFUL ENVIRONMENTS

The contributions of Bernard, Cannon, Selye, and others have been insightfully reviewed by many authors. Their publications provide a fresh perspective on several principles that are essential to a thorough understanding of the remaining chapters. For a more complete discussion, consult the suggested readings listed on page 323. Here are the basic principles you'll need to understand now:[4, 5]

1. Human cells, organs, and body systems are highly organized and are capable of undergoing change.

2. Each of earth's extreme environments requires a unique set of adaptive responses. But interestingly, adequate acclimitazation to virtually all of earth's stressful environments requires about 8-14 days of exposure, and the loss of acclimatization to these stressors (e.g., heat, cold, high altitude) occurs in about 14-28 days.

3. Humans may lack or have insufficient hereditary abilities to adapt to all of earth's stressful environments.

4. Stressful experiences result in personal growth, temporary perturbations, or permanent/deleterious effects. People vary in their capacity to respond to challenges, danger, and threats (see figure 1.1). The outcomes of stressful experiences are determined most by the factors listed in table 1.3 (page 6).

5. Different individuals respond to the same stressor with different outcomes. Their bodies may show no strain, illness, or injury, depending on the level of tolerance developed, immune system competence, age, level of physical fitness, and the number/intensity/type of previous exposure to this stressor.

Figure 1.1 One person's stress is another person's pleasure or leisure.

Table 1.3
Factors That Influence the Outcome of an Exposure to a Stressful Environment

Personal appraisal of the stressors (i.e., meaning)

Emotional response to the environmental situation

Genetic characteristics

Resources available

Age (especially in preadolescence and old age)

Nature and duration of previous exposures

Control or escape possibilities

Intelligence, education, skills

Number of similar previous experiences

Adapted from Weiner 1992.

6. Stressors may have positive effects that allow an individual to meet physical or psychosocial demands successfully. Among such effects are increased physical stamina and more effective coping styles.

7. Physical stressors (e.g., cold or heat exposure, sleep deprivation, prolonged or intense exercise) are strongly mediated by psychological factors. If these stressors are not viewed as noxious or alarming, they produce smaller or even opposite physiological responses.

8. Physiological responses at times can be excessive, inappropriate, inadequate, or disordered. These states are exemplified by chronic inflammatory diseases (e.g., rheumatoid arthritis), inappropriate **AVP** syndrome during water overconsumption (AVP also is known as ADH; it is a hormone that causes your kidneys to retain water), old age, and ventricular fibrillation, respectively.

9. If humans are able to prevent, avoid, control, or respond to a stressor, they usually will not become ill or injured.

10. Psychological strategies (e.g., coping maneuvers) may alter the amount of strain experienced by the individual when exposed to a stressful environment. This occurs because worry, fear, or panic result in strain that exceeds that from a stressful environment alone. These strategies include self-reassurance, self-deception, prayer, and seeking help from others.

11. **Mediators** are biological, social, and psychological modifiers that act on stressors to alter the level of physiological strain experi-

Spencer Swanger/Tom Stack & Associates

Mountain climbing often combines the extreme environments of high altitude and severe cold.

enced. A genetic deficit in a metabolic enzyme (biological), peer/parental expectations (social), and personality characteristics (psychological) are examples of mediators.

12. Finally, physiological and behavioral changes sometimes occur *before* a stressor is encountered, in anticipation of and in preparation for challenges, threats, and dangers. For example, athletes experience preevent jitters because the activity of the sympathetic nervous and the endocrine systems prepare the muscular and circulatory systems for competition. Even metabolism is enhanced in anticipation of a contest, as energy-rich compounds are transported to muscles. These interactions of the nervous and endocrine systems are explained in detail below.

YOUR BODY'S RESPONSES TO STRESS

Historically, three systems of the human body have been viewed as independent participants that are mobilized in response to stressors: the nervous, endocrine, and immune systems. In recent years, however, there has been a growing scientific awareness of interactions between these systems. In fact, for some purposes and responses, they may best be thought of as one large system serving an integrated function, rather than as three distinct systems. The paragraphs below consider each of these independently and present ways that they interact to provide a complex system of protection and adaptive responses.

Think back to a time that you exercised in a very hot, cold, or high-altitude environment. When you entered that extreme environment, your body had to sense the changes in its surroundings, as well as the change in homeostatic balance inside cells. Further, your body had to enhance specific physiological processes and coordinate responses between various systems that were appropriate to counteract the threats to the steady state of your internal environment. Only the central nervous system (CNS) (i.e., the brain, spine, and spinal nerves) could accomplish these tasks.

Shortly after your CNS sensed the external environmental stressors and the accompanying internal disruptions, the rate of release of certain hormones was altered by the CNS. This is especially true if you exercised intensely for a brief time, or moderately for a prolonged period. A **hormone** is a chemical that is released into the bloodstream by an endocrine gland and exerts its influence at a target organ. It is removed from the blood—that is, inactivated—after it has served its function. Both these events are regulated by the CNS and, interestingly, hormones influence the CNS in return. In some cases, a hormone causes the release of a second hormone, which then causes the release of a third hormone. Where such a sequence of hormone

© Jurgen Ankenbrand/J.A. Photographics

The Death Valley Badwater Marathon tests the limits of participants' adaptive responses.

actions occurs, we refer to the sequence of events as a **hormonal axis**. When you experienced this extreme environment, it is likely that two hormonal axes were activated: the *sympathetic-adrenal-medullary axis* (SAM) and the *hypothalamic-pituitary-adrenocortical axis* (HPA).[6]

Both the SAM axis and the HPA axis involve the adrenal glands, which are flat, caplike structures lying atop each kidney. Each adrenal gland consists of an outer portion (cortex) and an inner portion (medulla), as shown in figure 1.2. Because each part secretes different hormones, they are often considered to be two distinct glands. The **adrenal medulla** forms part of the SAM axis, and the **adrenal cortex** forms part of the HPA axis.

The medulla discharges its products (i.e., epinephrine and norepinephrine, which are part of the chemical family known as catecholamines) into the bloodstream and thus participates in the "fight or flight" reaction. Release of these catecholamines may be stimulated by low blood sugar, reduced blood volume or blood pressure, exercise, and stressors. The medulla is activated by the sympathetic branch of the autonomic nervous system; the parasympathetic branch (which serves to counteract the sympathetic) is generally quiet. Concurrently, some sympathetic nerves directly stimulate the release of norepinephrine in target organs and into the bloodstream (where it acts on distant organs) to contribute to the effects. As a result, blood pressure and cardiac output increase because blood is diverted to muscles from the stomach, intestines, liver, and kidneys. Other beneficial effects include increased glucose and free fatty acid mobilization in blood, increased cellular metabolism throughout the entire body, and increased mental activity.[7, 8]

The cortex produces, stores, and releases a group of hormones known as the corticosteroids, the most important of which is cortisol. Cortisol increases the production of carbohydrate from protein; increases the levels of glucose, protein, and fatty acids in blood; reduces inflammation; and suppresses the immune system.[7] The secretion of cortisol from the adrenal cortex is controlled by the secretion of the hormone ACTH (adrenocorticotropic hormone or corticotropin) in the anterior portion of the brain's pituitary gland. ACTH release initially is regulated by the secretion of CRH (corticotropin-releasing hormone) from the hypothalamus. Thus, the hypothalamus is involved in the translation of stressor-induced disruption of homeostasis in both the SAM and the HPA. Both external stressors and internal mental anxiety can rapidly increase the production of ACTH by as much as 20-fold.[7]

The distinct effects of these two hormonal axes complement each other. As figure 1.3 (page 11) illustrates, the primary hormone

The Body's Responses to Stress

Nervous System

Maintains homeostasis:

- brain receives afferent input
- brain sends efferent output that maintains homeostasis
- neurotransmitters (serotonin, dopamine, norepinephrine, GABA) are secreted and affect physiological responses (e.g., fuel mobilization, cardiovascular responsivity)

Endocrine System

- during stress, the catabolic hormones (cortisol, epinephrine, and norepinephrine) cause tissue breakdown, nutrient mobilization, and decreased cell metabolism
- recovery from stress increases anabolic hormones (insulin, testosterone, estrogen), enhancing tissue growth and increasing cell metabolism

Immune System

- immune function is enhanced by mild exercise training
- excessive stress causes immuno-suppression

All Other Systems

- all other body systems respond to various stressors in unique ways, to maintain homeostasis

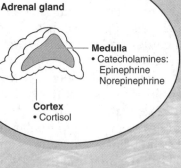

Brain

Adrenal glands

Kidneys

Adrenal gland

Medulla
• Catecholamines:
  Epinephrine
  Norepinephrine

Cortex
• Cortisol

Figure 1.2 Many body systems respond to stress.

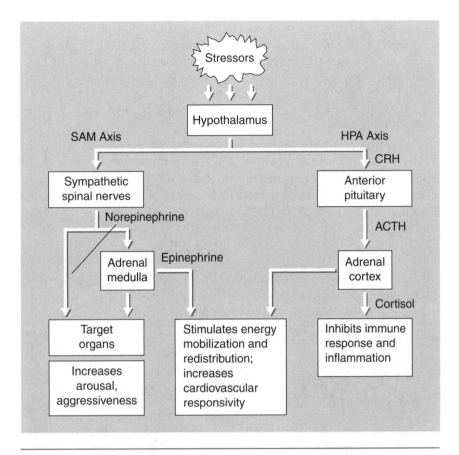

Figure 1.3 The two predominant hormonal axes involved in the body's response to stressors. CRH = Corticotropic-releasing hormone, ACTH = Adrenocorticotropic hormone.
Adapted from Berne and Levy 1996.

products of these axes—epinephrine, norepinephrine, and cortisol—all serve to mobilize and redistribute metabolic fuels (carbohydrate, fat, protein) at different rates and to enhance the responsiveness of the cardiovascular system by, for example, increasing the frequency and forcefulness of heart contractions. These responses prepare athletes and laborers alike for action. If environmental stressors cause tissue injury or trauma, high cortisol levels eventually act to restrain the initial inflammatory and immune responses so that they do not lead to permanent damage.[8]

Norepinephrine and CRH can produce other adaptive responses to stressors when they act locally on brain nerve cells as neurotransmitters. For example, their actions on the CNS result in a general state of arousal and aggressiveness. They also can inhibit behaviors such as growth, reproduction, ovulation, sexual activity, and eating,[8] which is necessary when the body's internal homeostatic balance has been disrupted. Considered collectively, these effects illustrate the integration between the nervous, endocrine, and immune systems.

Besides norepinephrine and CRH, at least five other chemicals involved in transmission of nerve impulses in the brain (i.e., neurochemicals) may play an integrating role in coordinating the body's responses to stressors.[6] Serotonin, for example, directly affects ACTH and prolactin secretion.[9] Dopamine (a catecholamine precursor of norepinephrine) release in the brain increases with exposure to various stressors.[10] The neurotransmitter GABA (gamma-aminobutyric acid) and endogenous opioids inhibit the function of the HPA axis.[9] Also, acetylcholine excites the release of CRH. These effects not only emphasize the complexity of the body's response to stressors, but also remind us that the SAM and HPA hormonal axes (see figure 1.3) are not the only pathways available.

Scientific observations also provide evidence for interactions between the nervous system and the immune system.[6, 11] These include the following:

Underwater welders experience extreme environments as part of their daily work.

- Animals exposed to stressors have altered immune status.
- Electrical stimulation of specific brain regions alters immune function.
- Changes in hormone and physiological exercise activities in brain cells are correlated with activation of the immune system.

In addition to the activities in the SAM and HPA hormonal axes, the secretion of other hormones changes when stressors are encountered.[12] These hormones can be classified as either anabolic (enhanced metabolism and tissue construction) or catabolic (decreased metabolism and tissue destruction). The former include insulin, testosterone, and estrogen; the latter include growth hormone, prolactin (the pregnancy hormone important for lactation) and thyroxine, in addition to epinephrine, norepinephrine, and cortisol. Interestingly, during a period of increased exposure to stressors, secretion of the catabolic hormones increases and the levels of the anabolic hormones in blood decrease. The end result, during stressful experiences, is a mobilization of energy reserves, as exhibited by growth hormone, prolactin, and thyroxine, which mobilize free fatty acids from adipose tissue. The poststressor outcome (when secretion of anabolic hormones increases and production of catabolic hormones decreases) builds depleted energy stores and replaces lost body tissue; for example, insulin promotes the storage of glucose and fats, and testosterone promotes protein synthesis.[6]

Table 1.4
The Effects of Intense, Brief Exercise and Moderate, Prolonged Exercise on the Secretion of Various Stress-Related Hormones

Hormone	Change in blood concentration	
	During intense, brief exercise	During moderate, prolonged exercise
Epinephrine	↑	—
Norepinephrine	—	↑
Cortisol	↑	—
Growth hormone	—	↑
Thyroxine	—	↑
Prolactin	—	↑

Symbols: ↑, increase; —, no change likely.
Adapted from McArdle, Katch, and Katch 1996.

Finally, the effects of stressful exercise on these hormones is especially pertinent to the chapters that follow because exercise throws the human body out of homeostatic balance even more than stressful environments alone. Table 1.4 summarizes these effects. Clearly, different stressor-induced hormones are enhanced by different types of exercise. During high-intensity exercise (i.e., power events involving predominantly anaerobic metabolism), epinephrine and cortisol are secreted. During endurance exercise (i.e., aerobic energy production), the production of norepinephrine, growth hormone, thyroxine, and prolactin increases. The text above describes the various influences that these hormones exert.

REFERENCES

1. Bernard, C. 1878. *Leçons sur les phénomènes de la vie communs aux animaux et aux vegetaux*. Paris: J.B. Bailliere et Fils.

2. Bernard, C. 1865. *Introduction à l'étude de la médecine expérimentale*. Paris: J.B. Bailliere et Fils.

3. Selye, H. 1956. *The stress of life*. New York: McGraw-Hill, 4-13.

4. Weiner, H. 1992. *Perturbing the organism. The biology of stressful experience*. Chicago: University of Chicago Press, 9-55.

5. Institute of Medicine, National Academy of Sciences. 1982. Conceptual issues in stress research. In *Stress and human health: Analysis and implications of research*, edited by G.R. Elliott & C. Eisdorfer. New York: Springer.

6. Toates, F. 1995. *Stress. Conceptual and biological aspects*. New York: Wiley, 2-6.

7. Guyton A.C., & Hall, J.E. 1996. *Textbook of medical physiology*. 9th ed. Philadelphia: Saunders.

8. Berne, R.M., & Levy, M.N. 1996. *Principles of physiology*. St. Louis: Mosby, 689-696.

9. DeSouza, E.B., & Appel, N.M. 1991. Distribution of brain and pituitary receptors involved in mediating stress responses. In *Stress—Neurobiology and neuroendocrinology*, edited by M.R. Brown, G.F. Koob, & C. Rivier, 91-117. New York: Marcel Dekker.

10. Dunn, A.J., & Berridge, C.W. 1990. Corticotropin-releasing factor as the mediator of stress responses. In *Psychobiology of stress*, edited by S. Puglisi & A. Oliverio, 81-93. Dordrecht: Kluwer.

11. Dunn, A.J. 1989. Psychoneuroimmunology for the psychoneuroendocrinologist: A review of animal studies of nervous system-immune system interactions. *Psychoneuroendocrinology* 14: 251-274.

12. Mason, J.W. 1972. Organization of psychoendocrine mechanisms. In *Handbook of psychophysiology*, edited by N.S. Greenfield & R.A. Sternbach, 3-91. New York: Holt, Rinehart & Winston.

chapter 2 Heat and Humidity

Years ago, when I was teaching high school biology and chemistry in the Midwest, I entered a 15K road race near West Unity, Ohio, on a sultry Saturday morning in July. At about the 8-K mark, I found myself cruising on a country road in the middle of a long procession of runners. Deep in thought and straining to maintain a fast pace, something caught my eye in a culvert that was recessed below the level of the road by about 1 m. It was a motionless body, lying face down. I knew that my race was finished. Within seconds, I realized that this was a race competitor who had been overcome by intense exercise in a hot-humid environment. It was exertional heatstroke!

I ran to a nearby house, summoned help, and called the local EMTs. We began pouring cold water on the body. Later, I learned that this unconscious runner was a student, working on his master's degree in exercise physiology.

That was the first time I realized that heat illness can strike anyone—even the well informed—if they push beyond their abilities in a hostile environment.

—Larry Armstrong

Chapter 1 explains ways that challenging environments and exercise can disrupt your body's homeostatic balance. Exercise in a hot environment provides a classic example of such an imbalance. Metabolic heat, produced during muscle contraction, must be removed from the central organs or they will overheat. Sweating causes a shift of internal body fluids, a reduced urine output, and usually leads to increased drinking. Circulatory and central nervous system adaptations maintain the blood's delivery of oxygen and essential nutrients to the brain, vital organs, and skeletal muscle.

Although the body can acclimatize to heat, if environmental and exercise stressors overwhelm the body's adaptive responses, heat illnesses occur. Because these illnesses involve imbalances of temperature, fluids, and electrolytes, they can be avoided. In this chapter, guidelines are provided that will help you make sound judgments, maintain good health, and optimize performance.

TEMPERATURE REGULATION

To maintain your temperature at about 37°C (98.6°F), your body is constantly adapting to changes in air temperature, humidity, air movement, solar radiant, barometric pressure, and clothing insulation. Further, as you metabolize food, approximately 80% of all energy in carbohydrates, fats, and proteins eventually is transformed to heat. Your body must remove this heat, or serious **hyperthermia** (high body temperature) could result. Environmental heat stress increases the requirements for sweating and circulatory responses to remove body heat. In addition, muscular exercise increases your metabolic rate above resting levels and increases the rate at which heat must be dissipated to the environment to keep your inner body temperature from rising to dangerous levels.[1]

Using the perspectives presented in chapter 1, heat and humidity can be viewed as unique stressors that impose an imbalance in temperature homeostasis. The resulting increase in body temperature is sensed by your brain, and increased sweat production is stimulated at many of the millions of sweat glands that lie just below your skin's surface. Concurrently, the brain causes the smooth muscles in your skin blood vessels to relax, allowing dilation and increased blood flow to the skin while the brain diverts blood away from inner organs. These are the primary means by which you dissipate heat. The evaporation of sweat cools the skin and increased skin blood flow carries heat from your body's core to the periphery, where warm skin releases heat to the air via nonevaporative (dry) means.

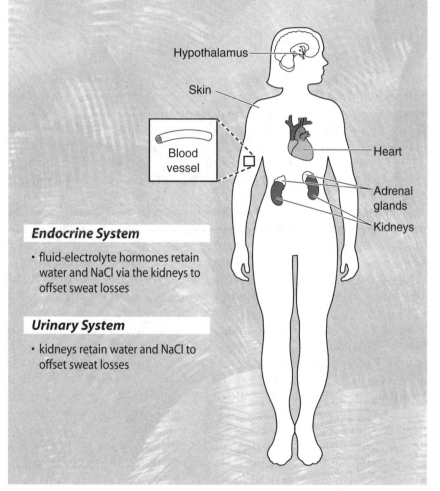

The Body's Responses to Stress

Nervous System

Maintains homeostasis:
- hypothalamus receives afferent input
- hypothalamus sends efferent messages that maintain/alter body temperature and regulate cardiovascular responses

Skin

- sweat glands secrete sweat to cool the skin

Cardiovascular System

- cutaneous blood vessels dilate, dissipating heat to the environment

Hypothalamus

Skin

Blood vessel

Heart

Adrenal glands

Kidneys

Endocrine System

- fluid-electrolyte hormones retain water and NaCl via the kidneys to offset sweat losses

Urinary System

- kidneys retain water and NaCl to offset sweat losses

Figure 2.1 Body systems and organs that respond to heat and humidity.

Figure 2.1 shows the primary organs that are involved in restoring elevated body temperature to about 37°C (98.6°F) through adaptive responses. Deep inside your brain lies a structure that has been identified as the region that senses environmental (skin) and deep body temperatures, integrates this information, and relays information to appropriate organs to initiate thermoregulatory responses. Known as the hypothalamus, this brain structure acts much like a room thermostat but also regulates thirst, fluid balance, hunger, metabolism, and reproductive hormones. In recent years, it has been recognized that the spinal cord, when heated, also influences temperature regulation by inducing responses in sweat glands and skin blood vessels.[2]

HOW YOUR BODY LOSES HEAT

Sweat originates deep in the secretory coil of the sweat gland and emerges at the skin surface to wet the skin. Sweat secretion over a given region of skin is dependent on both the density of sweat glands (i.e., number per cm^2) and the amount of sweat secreted per gland. In most people, the back and chest have the greatest sweating rates, while the arms and legs have relatively low sweating rates.[1] Evaporation of sweat occurs when water changes from a liquid to a gas. For this to happen, heat is supplied by skin. As skin loses heat, it is cooled. The necessary thermal energy equals 580 **kilocalories** (abbreviated **kcal**—the amount of heat needed to raise the temperature of 1 L of water 1°C) per liter of water. Therefore, evaporative cooling is most effective when the skin remains very wet, due to copious sweat production. High environmental temperatures and prolonged strenuous exercise may result in thermal sweating rates as high as 1.4-2.0 L/h.[3]

Sweat evaporation also is influenced by the amount of moisture in the air. Hot-dry air receives vaporized sweat readily. In contrast, hot-wet air receives little evaporated sweat because it is heavily laden with moisture. The meteorological measurement known as **relative humidity (rh)** provides an index of the amount of water in the air relative to a totally saturated volume of air (100% rh). As the relative humidity of air climbs over 50-70%, the effect on heat dissipation becomes obvious as more heat is stored in the body and is sensed as a hot, red skin. In a very humid environment, therefore, the body relies increasingly on nonevaporative dry heat loss via increased skin blood flow. This explains why your skin becomes red and flushed when you exercise in hot-wet conditions.

Figure 2.2 depicts the relative contributions of evaporative and nonevaporative heat loss, during rest and exercise, in both hot-wet and hot-dry environments. The evaporative heat loss due to sweat is abbreviated as E, whereas the total nonevaporative (dry) heat dissipation is abbreviated as R + C. This latter term refers to the combined effects of heat loss via radiation and convection. **Radiation** is the transfer of energy waves that are emitted by one object and absorbed by another. Solar energy from direct sunlight and radiant heat from the ground are examples. **Convection** is heat exchange that occurs between a solid medium and one that moves (i.e. a fluid); this movement is known as a convective current. Air and body fluids are technically considered fluids for this purpose. Conduction is not included in the figure because it accounts for less than 2% of heat loss in most situations. In convection and conduction, heat is transferred from a warm object to a cooler object.

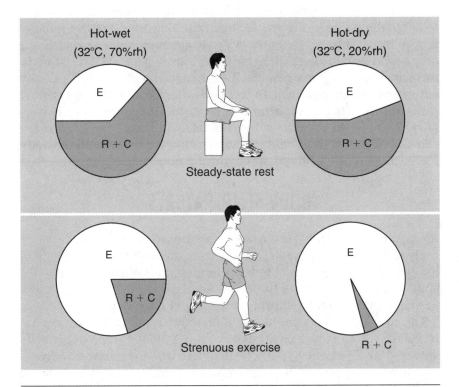

Figure 2.2 The relative contributions of radiation (R), convection (C), and evaporation (E) to total heat dissipation. Rest and strenuous exercise are shown, in hot-dry and hot-wet environments.

Adapted from Armstrong 1991.

As figure 2.2 shows, the body regulates its temperature differently in different environments and during different activities. In a hot-dry environment, evaporation accounts for 85-90% of all heat dissipation during exercise,[4] as depicted in the lower right quadrant of this figure, and emphasizes the need for wet skin and lightweight, loose-fitting, porous clothing.[5] However, in a hot-wet environment the evaporation of sweat from the skin surface is greatly diminished. The body must, as a result, rely more on radiative and convective heat loss. At rest, the ratio of E to R + C (see top half of figure 2.2) is similar in hot-wet and hot-dry environments. And, although it is not shown in this figure, R + C are major avenues of heat loss in a cool-dry environment, accounting for about 70% of such loss.[6]

Several other environmental factors are important in balancing body temperature. These include solar radiation, air speed, barometric pressure, and clothing insulation. However, few people recognize that R + C may not always act to remove heat from the body. In fact, the body gains heat from the environment when skin temperature is *less than* air temperature. If you realize that skin temperature usually ranges from 34 to 37°C (93 to 98°F), it becomes obvious that R + C may *add to* heat storage when air temperature nears 38°C (100°F). In this case, the skin *gains* heat via R + C. Unlike evaporation, R + C may either remove or add heat to the body, depending on the ambient temperature. This is important when both air temperature is high (>34-37°C, >93-98°F) and relative humidity is high (>50-70% rh) because heat loss via R + C and via evaporation are stifled, and the body stores most of the heat generated during exercise.

BODY TEMPERATURES

Rectal temperature (T_{re}) is the most widely used measurement of deep body temperature in medical, industrial, and athletic settings. Skin temperature (T_{sk}) represents the thermal state of the body's outer shell. Your brain controls these two areas separately, as it attempts to maintain thermal equilibrium throughout each day.

The combined effects of heat and humidity on deep body temperature during exercise are illustrated in figure 2.3. This graph depicts the steady-state T_{re} of one hypothetical athlete who exercised on a treadmill in many combinations of air temperature and humidity, at the same constant treadmill speed, on 27 separate occasions, until his body temperature reached a stable point (i.e., heat production equaled heat loss). The three symbols each represent a different relative humidity (either

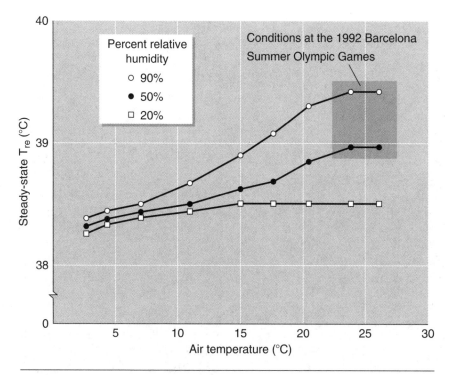

Figure 2.3 Steady-state T$_{re}$ of one hypothetical athlete who exercised in 27 different controlled environments. Symbols represent three different humidities.
Adapted from Armstrong 1992.

20%, 50%, or 90% rh). Clearly, increased humidity increases steady-state T$_{re}$ when air temperature is 16°C (60°F) or higher.[7]

One additional factor is important in determining the deep body temperature of an exercising human: **exercise intensity**. It should be obvious to you that you utilize food energy (carbohydrates, fats, proteins) at a faster rate during strenuous exercise than during mild exercise. This large amount of metabolic activity generates a great quantity of heat that must be removed from the muscles and other deep body tissues. Figure 2.4 (page 22) depicts the T$_{re}$ at nine different air temperatures for one hypothetical athlete. This athlete exercised on 27 different occasions, until his T$_{re}$ reached a plateau, at either 65%, 75%, or 90% of his maximal aerobic power, or $\dot{V}O_2\textbf{max}$. The three different symbols represent the different exercise intensities. Recognizing that the relative humidity was constant in all sessions, it is obvious that exercise intensity affects the final T$_{re}$, especially at ambient temperatures greater than 16°C (60°F).[7]

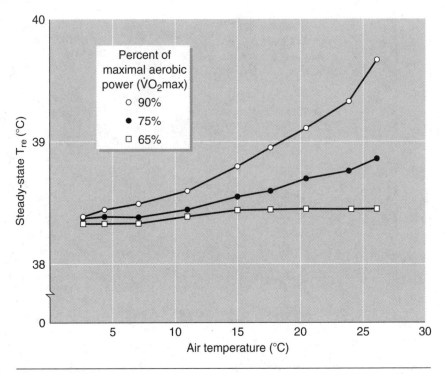

Figure 2.4 Steady-state T_{re} of one hypothetical athlete who exercised in nine different controlled environments, at three exercise intensities.
Adapted from Armstrong 1992.

CARDIOVASCULAR RESPONSES TO HEAT AND HUMIDITY

Your cardiovascular system (heart, blood vessels, and blood) is essential for temperature regulation and exercise performance in a hot environment. As noted earlier, the central nervous system distributes the blood ejected from the heart (cardiac output in liters per minute) to both skin blood vessels and active muscles. This is accomplished by increasing the cardiac output (from 6 L/min at rest to 25 L/min during strenuous exercise) by increasing heart rate and the volume of blood ejected with each beat (stroke volume). Further, the brain and nervous system divert blood temporarily away from inner organs such as the liver, intestines, and kidneys by constriction of the arteries that supply blood to them.[8] The result of this increased cardiac output and 30-40% reduction of blood flow to inner organs is that the cardiac output of the heart can be utilized for two important tasks: heat dissipation (increased skin blood flow) and exercise (increased muscle blood flow). Figure 2.5 illustrates this complex response as it is orchestrated by the brain and spinal cord. Constriction of arteries leading to inner organs is symbolized by C. Dilation of the arteries that supply skin and active muscle tissue is represented by the letter D.[8]

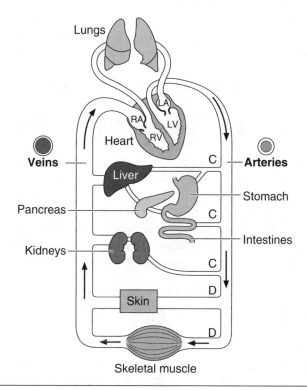

Figure 2.5 Control of the distribution of cardiac output to organs during exercise in the heat. C, constriction of arteries; D, dilation of arteries.
From *Human Circulation: Regulation During Physical Stress* by Loring B. Rowell. Copyright © 1986 by Oxford University Press, Inc. Used by permission of Oxford University Press, Inc.

Blood volume also plays an important role in the amount of physiological strain experienced during exercise in cool and hot environments. When humans move from rest to upright exercise, it is common to lose 5-10% of the fluid (i.e., plasma) contained in blood within the initial 5-10 min of activity. This plasma loss occurs because of the increased blood pressure that occurs during exercise. Compounding this loss of plasma volume is the loss of fluid in sweat. As noted earlier, body sweat losses can be considerable during prolonged exercise in summer months, sometimes reaching 6-10% of body weight.[9]

Such sweat losses may result in plasma volume decreases of up to 20%. This level of dehydration may be debilitating in hot-wet environments, when evaporation of sweat is inefficient and heat loss via R + C is required, because the brain increases skin blood flow, thereby decreasing the amount of venous blood that returns to the right atrium. This, in turn, causes insufficient filling of the heart and a reduced stroke volume with each heartbeat. To maintain cardiac output in this situation, the central nervous system increases heart rate, causing increased cardiovascular strain. Two of the common heat illnesses—heat exhaustion and heat syncope—result from this series of events; these illnesses will be described later in this chapter.

HEAT AND PHYSICAL PERFORMANCE

The question "What effects do heat and humidity have on exercise performance?" has no single answer for all types of exercise. Endurance and strength activities are affected differently because each involves a unique blend of heat exposure, exercise intensity, and opportunity for drinking fluids. In fact, from a physiological viewpoint, it is best to consider the influences of elevated body temperature (hyperthermia) and dehydration as the critical factors, rather than heat and humidity per se.

Hyperthermia

If you review figures 2.3 and 2.4, you will see the three most important causes of hyperthermia in athletes, soldiers, and laborers: high exercise intensity, high air temperature, and high relative humidity. This explains why *any athlete*, female or male, can experience performance-degrading hyperthermia if they push themselves too fast, for too long a time, in hot-humid conditions. There are three ways in which such hyperthermia might degrade physical performance:

1. Hyperthermia reduces **muscular endurance** (the ability to sustain muscular contractions for several minutes to hours). It may alter performance in the long-distance events of cycling, track and field, and soccer.[10, 11, 12] However, the peak force (maximal strength) exerted by muscles is virtually unaffected.[10, 12]

2. Hyperthermia shifts **metabolism** from primarily the aerobic (with oxygen) to the anaerobic (without oxygen) form. Unfortunately, this means that the body's stores of carbohydrate (e.g., glyco-

gen) in skeletal muscles and liver will be consumed at a faster rate.[13] This may partly explain why exercise in the heat cannot be maintained as long as in a cool environment.[14, 15] This metabolic response will affect events that rely heavily on carbohydrate stores in the body, such as endurance road cycling and marathon running.[7] In these contests, the body's limited carbohydrate stores can sustain no more than a couple of hours of moderate to intense exercise.

3. As noted previously (see the section titled "Cardiovascular Responses to Heat and Humidity"), hyperthermia causes the dilation of blood vessels in the skin and pooling of blood in the limbs. This reduces the volume of blood that returns to the heart, reduces cardiac output, and increases circulatory strain. These effects are perceived as increased fatigue[16] because the capacity to deliver oxygen to muscles is reduced.[17] All forms of prolonged labor and exercise may be affected by these thermoregulatory responses.

Dehydration

Although mild to moderate work and routine physical training typically result in whole-body sweat losses of 0.8-1.4 L/h, the highest sweating rate ever reported for an athlete was 3.7 L/h, in conjunction with the marathon of the 1984 Summer Olympic Games.[18] Further, we know that the maximum amount of fluid that empties from the stomach during exercise is 0.8-1.2 L/h in most athletes.[19] This explains why athletes routinely experience a 2-8% loss in body weight during competition and training.

Previous publications have examined the impact that dehydration has on muscular strength. These articles, written by respected physiologists, indicate that it is unlikely that small or moderate reductions in body weight due to dehydration (–1% to –3%) alter strength.[19, 20] In fact, dehydration to –5% or more can be tolerated without a loss of maximal strength. However, sustained or repeated exercise that lasts longer than 30 seconds deteriorates when moderate to severe dehydration exists (–6% or more).[21] This probably results from reduced muscle blood flow, waste removal, and heat dissipation—all of which are necessary for high-powered, sustained muscle action in events such as boxing, judo, and pursuit races in velodrome cycling.

In contrast, there is little doubt that acute dehydration, or its long-term counterpart chronic **hypohydration** (lasting 4 or more hours without rehydration), degrades endurance performance, regardless of the environmental temperature or whole-body hyperthermia. For example, maximal oxygen uptake ($\dot{V}O_2$max), a critical component of successful endurance performance, is reduced significantly following moderate body water losses (–3% of body weight) in a cool environment. In a hot environment, small to moderate levels of dehydration (–2% to –4%) result in a large $\dot{V}O_2$max decline. Similarly, endurance capacity (i.e., exercise time to exhaustion) is reduced more in a hot environment than in a cool or mild one.[20] Table 2.1 demonstrates the impact that dehydration has on endurance performance.[7] This summary of previous investigations shows that increasing dehydration (column 1) interacts with air temperature (column 2) to reduce both $\dot{V}O_2$max (a physiological measurement) and endurance capacity (a performance variable). You will recall from the previous section that weight losses of –2% to –5% had little effect on strength output. These observations coincide with the fact that decreases of plasma volume during prolonged exercise are greater in a hot (versus a mild) environment.[19] They also suggest that changes in cardiovascular function (e.g., decreased cardiac output) contribute to the decline in endurance capacity shown in table 2.1.

Table 2.1
The Impact of Dehydration on Endurance[8]

Body weight loss	Exercise environment	$\dot{V}O_2$max change	Endurance capacity change
–2%	Hot	–10%	–22%
–4%	Hot	–27%	–48%
–5%	Mild	–7%	–12%
–5%	Mild	——	–17%

Adapted from Armstrong 1992.

COMBATING HYPERTHERMIA AND DEHYDRATION

In light of the facts above, every athlete, laborer, or soldier should develop a plan to minimize the stress of living, working, training, and competing in hot-humid conditions. If this is not done, decrements in physical performance will be inevitable. The focus of this plan should be to minimize the effects of hyperthermia and dehydration.

Heat acclimatization greatly enhances contestants' chances of completing the Marathon de Sables—a 150-mile foot race across the Sahara Desert.

Heat Acclimatization

Prior to strenuous exercise or labor in a hot environment, the most important preventive measure involves exposing oneself to exercise-heat stress gradually, on consecutive days. This process, known as **heat acclimatization (HA)**, stimulates adaptive responses that improve exercise performance and heat tolerance, and reduce physiological strain and the incidence of some forms of heat illness. These adaptive responses include decreased rectal temperature (T_{re}), heart rate (HR), and psychological perception of effort, plus increased exercise, tolerance time, plasma volume, and sometimes increased sweating rate.[22] It also has been reported that increased sweat sensitivity (sweat production expressed per degree rise of T_{re}) and decreased sodium chloride (table salt; the chemical abbreviation is NaCl) losses in sweat and urine occur during HA, but these adaptations are dependent on environmental conditions[23] and sodium levels in the diet.[24] Table 2.2 (page 28) can be used to identify the organs and systems that adapt during HA. These include the heart, blood, sweat glands, kidney, hormones, and temperature regulatory organs (figure

Table 2.2
"Plateau days" of Physiological Adaptations (Point at Which Approximately 95% of the Adaptation Occurs) During Heat Acclimatization

Adaptation	Days of heat acclimatization														
	1	2	3	4	5	6	7	8	9	10	11	12	13	14
Heart rate decrease			—	—	—									
Plasma volume expansion			—	—	—									
Rectal temperature decrease				—	—	—	—							
Perceived exertion decrease			—	—	—									
Sweat Na$^+$ and Cl$^-$ concentration decrease*					—	—	—	—	—					
Sweat rate increase							—	—	—	—	—	—	—	
Renal Na$^+$ and Cl$^-$ concentration decrease			—	—	—	—								

*While consuming a diet low in NaCl.

Reprinted from Armstrong and Dziados 1986.

2.1). The results of these physiological adaptations, explored in detail below, are (a) improved heat transfer from the body's core to the skin, and ultimately to the external environment, and (b) improved cardiovascular function that copes with the dual stressors of dehydration and decreased blood volume (to inner organs) due to increased skin blood flow.

Table 2.2 summarizes the results of numerous HA investigations published during the last 50 years. The horizontal bars represent the range of days required for each specific adaptation to achieve approximately 95% of its maximal response in healthy, well-nourished, adequately hydrated adults.[25] Four concepts emerge from this table.[22] First, HA is a complex of adaptations, involving many organs and systems of the body, that serve to enhance heat dissipation and reduce cardiovascular strain. Second, the systems of the human body adapt to successive days of exercise-heat exposure at varying rates. Third, this process requires up to 14 days before complete acclimatization is achieved in all systems and organs. Fourth, the *early* HA adaptations primarily involve an improved control of cardiovascular function, due to expanded plasma volume, reduced heart rate, and autonomic nervous system adaptations (i.e., habituation), which redirects cardiac output more effectively to skin capillary beds and active muscles. The plasma volume expansion is an extremely important early adaptation, which ranges from +3% to +27%[24] and results in a 15-25% decrease in heart rate.[26] This occurs as a result of a plasma protein increase, causing retention of water in the circulation. It apparently is a temporary response, which decays after 8-14 days, as do the fluid-regulatory hormone responses,[27] and then is replaced by later, long-lasting adaptations such as increased sweat rate and diminished skin blood flow.[22, 28] The sweat and urine NaCl conservation shown in table 2.2 (via the hormone aldosterone) also contribute to the expansion of plasma volume, by increasing the total volume of extracellular fluid. This, in effect, reduces cardiovascular strain despite ongoing dehydration and competition for cardiac output between contracting muscles and skin capillary beds.

The following recommendations are offered to assist you in optimizing your HA efforts:[7]

1. You can facilitate heat acclimatization by exercising in the heat at intensities greater than 50% $\dot{V}O_2$max. Ideally, your total exercise-heat exposure time each day should be 90-100 min, but you should reach this duration gradually by increasing duration and intensity during the first 10-14 days of hot weather training.

2. Exercise with a partner during the initial days of HA training and on very hot days.

3. When moving from a cool to a hot environment, or when moving from a hot-dry to a hot-wet environment, you will find it difficult to perform strenuous workouts that you previously completed in less stressful surroundings. Rather than accept a slower pace in the heat, it is preferable that your high-intensity training be conducted in the cooler air of early morning to allow muscles and other organs to experience rigorous training.

4. Monitor T_{re} during or immediately after HA training, to ensure that your body temperature remains within safe limits (less than 39°C, 102°F). Use a glass rectal thermometer, purchased at a pharmacy, that reads high body temperatures (over 40°C, 104°F).

5. You can partially accomplish HA by exercising in a heated room or wearing insulated clothing, but do not stifle heat dissipation totally. Take care to avoid extreme hyperthermia and heat illness by measuring T_{re}.

Acute Changes in Body Water

In a mild environment, the average adult gains and loses about 2.5 L of water each day. The majority of fluid loss in this situation is due to urine, but feces, respiratory water, and water diffusion through skin (other than sweat) also contribute. In a hot environment, most fluid is lost as sweat, which may exceed 10 L if heat exposure lasts for an entire day. Such long-term heat exposure, without air-conditioning, may result in gradual dehydration, heat illness (see below), or changes in disposition. Because the rate of dehydration is slow, these effects require several hours to develop.

It is important for athletes, and anyone who lives or works continuously in a hot environment, to remain aware of dehydration as it develops. Perhaps the simplest way to accomplish this is to measure body weight before and after an activity or lengthy exposure to heat. A decrease in body weight indicates that body water has been lost; very little of this body weight change will be due to fat loss or other tissue breakdown. If you follow the simple steps in figure 2.6, you will be able to calculate your hourly sweat rate. Once you know your hourly rate of sweat loss, you will have a handy fluid replacement guide to use whenever you exercise or live in the heat. Remember, however, that sweat rate will change as body temperature, activities, exercise intensities, environmental conditions, and your heat acclimatization status change.

1. Record body weight before and after 1 h of exercise or rest in heat.
2. Calculate the difference (D) between these weights.
3. If clothing is wet, the increase in clothing weight should be noted by weighing it before and after heat exposure. This increase in weight represents sweat that was produced but remained in the fabric, and should be *added* to D (step 2 above). If nude body weight is taken, this step can be ignored.
4. The weight of fluid consumed during step 1 should be *added* to D.
5. The weight of urine lost during step 1 should be *subtracted* from D.
6. Your hourly sweat rate (in kg/h) equals D (step 2), after correcting for the items in steps 3-5.
7. Weight loss should be replaced by consuming 1 L of fluid for each kilogram of body weight that was lost.

Sample calculation
1. Clothed body weight before exercise = 70.0 kg. Perform 1 h of mild exercise in a hot environment. Clothed body weight after exercise = 69.0 kg.
2. D = 70.0 − 69.0 = 1.0 kg.
3. Clothing weight before exercise = 1.0 kg. Clothing weight after exercise = 1.5 kg. Increase in clothing weight due to sweat in fabric = 0.5 kg.
4. Fluid consumed during exercise = 0.5 L = 0.5 kg.
5. No urine was lost.
6. D = 1.0 kg (step 2) + 0.5 kg (step 3) + 0.5 kg (step 4) = 2.0 kg.
7. Weight loss should be replaced by consuming: (1 L/kg) × (2.0 kg) = 2 L of fluid.

Figure 2.6 Calculate your hourly sweat rate.

The American College of Sports Medicine (ACSM), the world's foremost professional organization of its kind, has published comprehensive recommendations regarding dehydration, athletic performance, and fluid losses. In the 1996 ACSM Position Stand "Exercise and Fluid Replacement" (see appendix A), the following guidelines are offered regarding the optimization of exercise performance:[29]

- At least 2 h prior to exercise, you should consume 500 ml (about 17 oz) of fluid; this promotes proper hydration and allows time for the kidneys to excrete excess water, if any exists.
- During exercise, you should begin drinking early, before you become thirsty. You should set a goal of consuming fluids at a rate equal to your sweating rate (see figure 2.6), or at the maximal rate that can be tolerated (600-1200 ml/h, or about 20-40 oz/h).
- Fluids should be cool (15-22°C, 59-72°F), palatable, and readily available in ample volumes.

Are Fluid Replacement Beverages Superior to Water?

For all practical purposes, fluid replacement beverages contain three ingredients that may assist exercise performance: water, carbohydrates, and salts (e.g., electrolytes or minerals). Thus, it is pertinent to ask the question, "Are fluid replacement beverages superior to plain water when consumed during and after exercise?" Recent research conducted in our laboratory at the University of Connecticut provides some insight.[30] In this investigation, water replacement per se was responsible for reversing cardiovascular and thermal strain, due to a 3.5% body-weight loss, during 90 min of treadmill exercise in a warm environment (33°C, 91°F). This occurred irrespective of carbohydrates and salts and suggested that much of the beneficial effect of fluid replacement beverages is due to water alone. In fact, our recent review of fluid consumption during exercise noted that there are four situations in which consuming commercial beverages appear to be superior to pure water. First, if a carbohydrate deficiency exists in blood, liver, or muscle tissue. This most often occurs during single exercise bouts lasting longer than 1 h. Second, when exercise is strenuous (>70% $\dot{V}O_2$max) and lasts longer than 50-60 min. There is very little evidence that carbohydrates in fluids affect performance when exercise lasts less than 50 or 60 min, regardless of its intensity. Third, if a sodium or NaCl deficiency exists. In most cases, sodium lost in sweat can be replaced with normal dietary intake.[29] Fourth, if it is necessary to *rapidly* replace lost plasma water after exercise, at a rate faster than that provided by normal meals. Other than these circumstances, there is

© Kristin Olenick/New England Stock

In most instances, water is the only fluid replacement you need.

no definitive evidence that carbohydrate-electrolyte beverages enhance exercise performance in excess of the effects derived from consuming an equal volume of pure water and a balanced diet.[29, 31]

This does not mean that fluid and carbohydrate replacement have no positive effects on exercise performance. They usually do during lengthy endurance efforts such as cycling and running. The primary role of water is to offset the detrimental effects of dehydration on endurance performance. The role of carbohydrates is to maintain blood glucose, enhance carbohydrate metabolism, and delay fatigue, especially when muscle glycogen is low. To maintain blood glucose, carbohydrates should be ingested throughout exercise by consuming beverages with less than 10% carbohydrate content (10 g per 100 ml of fluid). For example, if the desired volume of fluid is 0.6 to 1.2 L (20-40 oz) per h, this requirement can be met by drinking beverages in the range of 4-8% carbohydrate.[29] However, two clarifications of this data must be made. The first involves the scientific literature regarding carbohydrates and exercise performance published between 1979 and 1995. Sixty-two percent of all trials resulted in no significant statistical difference between pure water and fluids containing carbohydrates.[31]

This indicates that it is unlikely that carbohydrate consumption will enhance exercise performance in all recreational and athletic situations. The second clarification involves the groups of people that fluid replacement beverages will most likely benefit. Because only a small portion of adults around the world can exercise for 50-60 min strenuously (>70% $\dot{V}O_2$max), the vast majority of adults will not benefit from the exercise-enhancing effects of carbohydrate-containing fluids. Further, very few adults, during routine weekly training, ever experience a serious NaCl deficit or need to expand plasma volume rapidly after exercise. Therefore, the vast majority of adults will replace salt lost in sweat by consuming a normal diet, which is usually high in NaCl. In fact, the greatest need for carbohydrate-electrolyte replacement fluids occurs among athletes, soldiers, or laborers who (a) lose more than 7.6 L (8 qt) of sweat per day; (b) are not heat acclimatized; (c) skip meals, have meals interrupted, or encounter a heat-induced loss of appetite; (d) experience a caloric deficit of >1000 kcal/day; or (e) are ill with diarrheal disease.[32]

Chronic Changes in Body Water Across Days

Athletes and laborers should avoid beginning a training session or competition with a body water deficit that originated on the previous day. However, it is difficult to know how much fluid is needed, if any, unless accurate body-weight measurements are taken at the same time each day (such as in the morning after waking), while wearing the same clothing. It also is wise to use a sensitive digital scale that employs a strain gauge, not a spring mechanism, to measure body weight. If day-to-day changes in body water are detected, corrective steps should be taken that are dictated by the extent of the hypohydration.[7] For example, athletes reporting to a training session 2-3% lighter than on the previous day should rehydrate prior to exercising, by following the preexercise recommendations of the ACSM (see the bulleted list on page 31). Athletes reporting with a 3-6% body-weight deficit should reduce training intensity and duration that day and concentrate on replacing lost fluids. If the body-weight deficit is 7% or more, the athlete should consult with a sports medicine physician for a medical examination and consultation.[15, 22]

Urine: An Index of Fluid Balance

The kidneys regulate fluid and electrolyte balance very accurately from one day to the next. Even though strenuous, prolonged exercise

(1-2 h) might result in large sweat losses, the balance of water and salt is restored within hours, for most athletes, if they consume a normal diet. This is regulated by two hormones: aldosterone, which controls sodium and chloride balance, and **AVP** (arginine vasopressin, also known as ADH, or antidiuretic hormone), which controls water balance. Unless sweating causes a loss of body water that exceeds 3% of body weight, aldosterone and AVP regulate whole-body fluid and electrolyte balance within 1% of normal levels, on any given day.

Knowing this, it is reasonable to expect the properties of urine to reflect body water status. That is, urine should be concentrated and scanty when the body is dehydrated and is conserving water. The opposite ought to be true as well; urine should be dilute and plentiful when a temporary excess of body water exists. Indeed, this is precisely what occurs. For example, the average adult excretes approximately 1.2 L each day. If your urine volume is less than 1 L during any 24 h period, your body is conserving water via the hormone AVP. This is a sign that you should consume extra fluids during and between meals. This also explains why nutritionists and exercise physiologists recommend checking urine volume as one means of determining hydration status.

Specific Gravity Relevant laboratory techniques that analyze urine properties include measuring either osmolality or specific gravity. **Osmolality** refers to the concentration of a sample and is affected by all dissolved particles in a standardized volume (e.g., mass) of fluid. Measurements of osmolality require a sophisticated instrument, a trained laboratory technician, and are time-consuming. **Specific gravity** refers to the density (mass per volume) of a urine sample in comparison to pure water. Any fluid that is denser than water has a specific gravity greater than 1.000, and normal urine specimens usually range from 1.013 to 1.029 in healthy adults. During dehydration or hypohydration, urine specific gravity exceeds 1.030. When excess water exists, values from 1.001 to 1.012 are typically seen.[33] Specific gravity can be measured quickly and accurately with a handheld device known as a refractometer, shown in figure 2.7. First, one or two drops of a urine specimen are placed on the stage of the refractometer (see figure 2.7a). Next, this instrument is held up to a bright light, which passes through the specimen and through a lens, causing the specific gravity to appear on a scale that ranges from 1.000 to 1.040 (figure 2.7b). Due to its ease of operation, this device can be used indoors or outdoors. Available from scientific supply companies, the

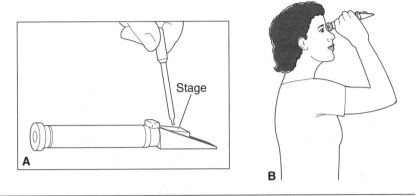

Figure 2.7 Use of a refractometer to measure urine specific gravity.

cost of inexpensive models begins at $175. Despite its relative ease of use, purchasing and using a refractometer still may be too intimidating for the average athlete.

Urine Color In an attempt to simplify the analysis of urine, with a method that all athletes might use, our research team has conducted a series of experiments involving the color of urine (U_{col}). We reasoned that if U_{col} were directly proportional to the level of dehydration, virtually anyone could determine when they need to rehydrate. We began by developing a U_{col} scale that could be used to derive a number or rating. This color scale was based on previous observations of urine samples, collected in field and laboratory studies of soldiers, dating back to 1988, and was printed on a laminated chart. Since publishing our findings regarding this U_{col} scale in 1994, we have had numerous requests from nutritionists, coaches, and physiologists around the world.[33] Your copy appears on the rear cover of this book. As you can see, the eight-color scale includes colors ranging from very pale yellow (number 1) to brownish green (number 8).

We were very excited to learn that this color scale validly reflects body hydration status. Our initial publication, for example, demonstrated that U_{col} can be used in field settings where close estimates of urine osmolality and specific gravity are acceptable, or when laboratory measurements are not practical.[33] Recognizing the limitations of color names (e.g., the meaning of "yellow" is different for each person), we recommend that athletes seek to consume water and other fluids to the point that urine is either "very pale yellow," "pale yellow," or "straw colored." These terms correspond to the numbers 1, 2, and 3, respectively, on the U_{col} scale. If this is accomplished, athletes will be well hydrated and urine specific gravity will be less than 1.014. Our initial publication provided two additional insights.[33] First, some individuals may require more than verbal instructions before they learn how much fluid must be drunk to produce dilute urine. This became obvious when a few subjects arrived at the laboratory in a hypohydrated state, despite instructions to drink additional water. Second, the day-to-day reliability of U_{col} measurements is enhanced if meals, fluid consumption during exercise, the time of urine collection, and training are consistent.

Our second publication was prompted by a field study in which our urine color chart was used by members of the British swimming team that competed in Atlanta at the 1996 Summer Olympic Games.[34] Neal Pollock, then affiliated with Florida State University in Tallahassee, followed these athletes for up to 9 days prior to competition. Professor Pollock's observations (unpublished, 1996) suggested that U_{col}, urine specific gravity, and urine osmolality were not as strongly correlated with dehydration as our initial observations had indicated. We believed that periods of heavy physical training and large water turnover were responsible, and designed a study to test this hypothesis. Its purpose was to evaluate the sensitivity with which urine properties reflected body water changes during marked dehydration (–4% of body weight), a strenuous exercise trial in a hot environment (37°C, 98.1°F), and a 21 h period of oral rehydration. The change in body mass (i.e., weight expressed in kilograms) was used as the "gold standard" by which all hydration indices were evaluated, because it represented body water fluctuations.

Figure 2.8 presents four graphs that compare changes in body mass, U_{col}, urine specific gravity, and urine osmolality for highly trained cyclists, across the five phases of this study: B, baseline state, before testing; D, dehydration; E, exercise to exhaustion; 4 h, 4 hours of ad libitum rehydration; and 21 h, 21 hours of ad libitum rehydration. Each point represents the average value of nine test subjects; the vertical bars represent the amount of variability (i.e., standard deviation) that occurred at each time point. These graphs clearly show that U_{col}, specific gravity, and osmolality followed a pattern similar to that of fluid loss. The minor exceptions to this pattern, in urine specific gravity and urine osmolality at phase D, indicate that U_{col} (measured with our eight-color scale) mimics body water loss as effectively as, or more effectively than, urine specific gravity or osmolality. And, although it is widely acknowledged that U_{col} changes in response to illness, medications, some vitamin supplements, and food pigments,[35] such effects

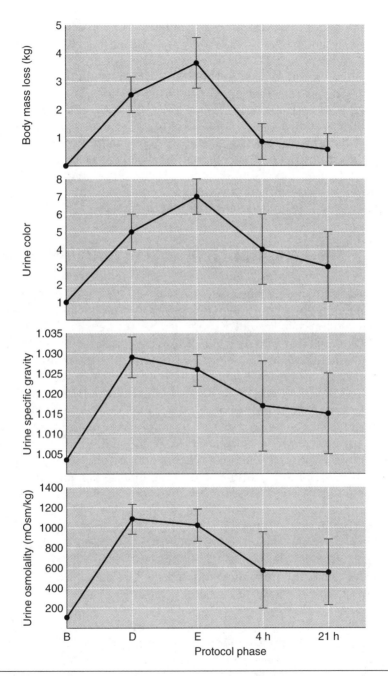

Figure 2.8 Changes of body mass, urine color, urine specific gravity, and urine osmolality during the baseline period (B), dehydration (D), exercise (E), and rehydration of 4 hours (4 h) and 21 hours (21 h).

Reprinted from Armstrong 1998.

were minimal in this study. We concluded that a 4% dehydration and lengthy, sometimes rapid, rehydration had little effect on the ability of U_{col} to mimic changes in body water. It also was interesting that U_{col} tracked body water fluctuations better than urine volume (not shown in figure 2.8). One exception was noted, however. This involves the times when dehydrated athletes rapidly rehydrate with a large quantity of pure water or dilute fluid, resulting in one or two urine specimens with a low U_{col}, specific gravity, or osmolality, before the body water deficit has been replaced completely. In cases such as this, especially when no food is consumed, whole-body rehydration can be monitored by observing three or more consecutive samples that indicate that urine is dilute, during a period of 3-5 h.

THE IMPORTANCE OF DIETARY SODIUM

This chapter has described several ways that proper fluid-electrolyte balance helps to optimize physical performance and reduce the risk of heat illness. In fact, all of the heat illnesses below may be treated with either oral or IV fluids (table 2.3). Therefore, it is reasonable to ask the question, "How much salt is required to maintain physical performance and health?" and "Do active people, who work or compete in hot environments, require more salt than sedentary individuals?" The paragraphs below focus on sodium, one of the elements in table salt (NaCl), because sodium is considered to be the body's most important electrolyte. It is involved in the regulation of water movement between the intracellular and extracellular compartments, nerve conduction, cellular metabolism, and the maintenance of blood volume, osmolarity, and pressure. Because sodium is constantly being lost from the kidneys, eccrine sweat glands, salivary glands, and gastrointestinal tract, and because sodium is consumed in most foods and beverages, the human body must regulate the sodium concentration of all body fluids closely. As noted earlier, if plasma sodium levels decrease, the hormone aldosterone is released from the adrenal cortex, and the loss of sodium is reduced in urine, sweat, and saliva.

Prior to World War II, supplemental dietary sodium intake was viewed as an asset by the medical community and media.[36] Citizens were advised to increase their consumption of NaCl during periods of hot weather; salt was used extensively as a preservative for foods, and it was added in large amounts to many foods (e.g., peanuts, pickles, pretzels, potato chips) to enhance their flavor. After World War II,

Table 2.3

The Four Heat Illnesses That Occur Among Athletes and Laborers

	Diagnosis	Treatment
Heat exhaustion	Inability to continue exercise in the heat. Rectal temperature 39°C (102°F), depending on the physical activity that preceded overt illnesses and the point at which temperature was first recorded. Sweating is profuse. Mental function and thermoregulation are mildly impaired. Acclimatization reduces the incidence of symptoms. See table 2.2, page 28.	Rest and cooling increase venous blood flow to the heart. Typical losses during a 4 h work shift in harsh conditions are 6.0 L water and 8 g NaCl. Replace these orally or with IV fluids. The prognosis is best when mental acuity is not altered and when serum enzymes are not elevated. Immediate return to work/exercise is not advisable except in the mildest cases; allow 24-48 h for recovery. Serious complications are very rare, unless prolonged hyperthermia is involved.
Exertional heatstroke	Thermoregulatory overload or failure. Rectal temperature of 40°C (104°F) or higher. Other symptoms include elevated serum enzymes, vomiting, diarrhea, coma, convulsions, and impairment of mental function and temperature regulation. Sweating may or may not be present. Onset may be rapid in patients who have been exercising. See table 2.7, page 52.	A true medical emergency. Water immersion not only provides the fastest cooling rate when rectal temperature is >40.6°C (>105°F) but also improves venous return and cardiac output via skin vasoconstriction and the effects of hydrostatic pressure. The mortality rate (10-80%) is directly related to the duration and intensity of hyperthermia as well as the speed and effectiveness of diagnosis and whole-body cooling.

(continued)

Table 2.3
(continued)

	Diagnosis	Treatment
Heat cramps	Associated with whole-body salt deficiency. Cramps occur in the abdominal and large muscles of the extremities but differ from exertion-induced cramps since the entire muscle is not involved; cramp appears to wander because individual motor units contract. Plasma Na^+ deficit with urine specific gravity >1.016. Observed mostly in unacclimatized individuals.	Oral 0.1% saline solution (two 10-grain salt tablets or 1/3 tsp of table salt in 1 L of water) or IV normal saline solution. IV solutions, used when symptoms include nausea and vomiting, bring rapid relief with no lasting complications.
Heat syncope	Brief fainting spell without a significant increase in rectal temperature. Pale skin is obvious. Pulse and breathing rates are slow. Presyncope warning signals include weakness, vertigo, nausea, or tunnel vision.	Lay the patient in the shade and elevate feet above the level of the head. Replace fluid and salt losses. Avoid sudden or prolonged standing.

Abbreviations: IV = intravenous, g = gram, Na^+ = sodium, NaCl = sodium chloride (table salt), tsp = teaspoon.

Adapted, by permission, from L.E. Armstrong and R.W. Hubbard, 1989, "Hyperthermia: New thoughts on an old problem," *The Physician and Sportsmedicine* 17(6):98.

sodium lost its favored position among medical professionals, because it was recognized as a key etiological factor in the development of salt-sensitive hypertension. In 1977, a select committee of the U.S. Senate (McGovern Committee on Nutrition and Human Needs) recommended that U.S. residents reduce their salt consumption to about 3 g NaCl/day. Currently, the Food and Nutrition Board of the National Research Council recommends that adults limit their daily intake of sodium to 6 g NaCl/day.[37]

The exact amount of sodium required by humans, at any stage of development or during any specific activity, is difficult to measure because the absolute need is determined almost entirely by nonrenal losses such as sweating. Despite this fact, investigations have shown that the basal metabolic requirement for salt is remarkably small. For example, Dahl reported that hospitalized patients required only 0.1-0.75 g NaCl/day.[38] Except for a few cases of renal impairment, maintenance of sodium balance and health was not a problem. Clearly, the average daily salt intake of U.S. adults (5.8-17.5 g NaCl) far exceeds their physiological needs.

It also is difficult to determine the sodium requirements of individuals who leave modern lifestyles in mild climates and move to hot climates. Many factors affect sodium turnover, including sweat production, state of heat acclimatization, dietary intake of sodium, and daily duration of exposure to heat. One analytical method involves measuring the dietary salt levels of healthy residents of hot environments. For example, studies have shown that the Masai warriors of Africa (<5 g NaCl/day), and Galilean Naturalists who eat no livestock (1.9 g NaCl/day), maintain year-round health on low-salt diets.[14] These studies support Dahl's findings,[38] and suggest that extreme hormonal sparing of sodium in sweat and urine, controlled water intake, careful dietary practices, as well as clothing may affect sodium balance in a hot climate. Because many diverse populations have been studied, it is unlikely that there are major hereditary influences on basal sodium needs.

In contrast to these studies, many experts formerly recommended large sodium intakes. These were published between 1938 and 1966, and involved recommended daily levels of 15-48 g NaCl for continuous desert living. Hubbard and colleagues[39] reviewed these studies and concluded that such recommendations were unnecessarily large, in light of aldosterone-mediated sodium conservation that occurs during heat acclimatization. They noted that excess dietary salt results in reduced plasma aldosterone levels, exactly opposite the hormonal status desired during prolonged thermal stress. This is especially true if secondary challenges (e.g., reduced food intake, occupational distractions, increased work requirements) occur concurrently.

Laboratory studies indicate that low dietary sodium intakes are safe during successive days of prolonged exercise-heat exposure.[40, 41, 42] The protocols of such studies involved controlled low-sodium diets (4-6 g NaCl/day), hot environments (>40°C, 104°F), prolonged daily exercise (1.5-4 h/day), and large sweat losses (1-6 L/day) that altered sodium balance. The findings of these studies were as follows: (a) sodium homeostasis was accomplished successfully while consuming a low-sodium diet after 8-10 days of exercise in heat; (b) the risk of heat illness (e.g., heat exhaustion) was greater on days 1-3, but no cases of heat illness were observed once sodium balance had been reached; (c) hourly replacement of fluid losses (due to sweat and urine) with an equal volume of pure water was important in the development of heat acclimatization and the maintenance of exercise performance; and (d) the following diets resulted in similar physiological responses, 4 g versus 8 g, 6 g versus 29 g, and 5.7 g versus 23.2 g NaCl/day.

When the kidneys regulate sodium reabsorption, potassium is exchanged for sodium and lost in urine. This process was previously proposed as a mechanism for the potassium depletion observed in permanent residents of hot environments and in some soldiers undergoing basic training. The clinical signs and symptoms of intracellular potassium depletion include impaired nerve conduction, muscular weakness, and cardiac rhythm abnormalities.[14] Although potassium depletion may result from chronic exposure to a hot environment, it is unlikely in the vast majority of athletic and industrial situations.

A low extracellular sodium concentration may be observed in a small percentage of endurance athletes. This condition, known as **hyponatremia**, involves a plasma sodium concentration below 130 mEq Na^+/L. Although athletes typically encounter this illness subsequent to prolonged competition lasting more than 7 h, hyponatremia has been observed during exercise of only 4-5 h duration. Four theoretical causes exist:[43] (1) sodium losses and plasma sodium levels are normal, but a large volume of pure water or hypotonic fluid is consumed and retained, thereby reducing the concentration of extracellular sodium; (2) sweat volume and sodium concentration are very great, resulting in sodium depletion; (3) a large, inappropriate secretion of the hormone arginine vasopressin induces excessive free water retention; and (4) etiologies 1-3 combine to regulate extracellular fluid tonicity and volume inappropriately. Although proponents may be found for each of these mechanisms, recent evidence supports the first etiology because the ingestion of excessive volumes of hypotonic fluid (e.g., 10 L in 4 h) during exercise has led to hyponatremia[43, 44] and the following signs and symptoms: increased body mass, extreme fatigue, nausea, and disorientation. Severe cases of water intoxication also have involved combinations of grand mal seizure, pulmonary edema, increased intracranial pressure, and respiratory arrest. It is probable that an overemphasis on drinking requirements leads to this syndrome. In most training or competitive situations, a weight gain indicates that excess fluid has been retained in the body. Because fluid replacement is necessary for health and maintenance of performance in the vast majority of exercise scenarios, athletes must learn to ingest reasonable, not excessive, quantities of fluid.

As noted above, the average daily diet of modern cultures provides more sodium than exercise-induced sweating usually demands; the excess sodium is excreted by the kidney. It is only when humans dramatically depart from normal routines (e.g., athletes participating in ultraendurance events, extended heat exposure while not acclimatized) that deficiencies occur. Therefore, the key to avoiding all sodium imbalances that impart decrements in performance or illness is to match sodium and water consumption with acute sodium and water losses. You can accomplish this in three ways. First, you should eat normal meals and neither restrict nor greatly increase sodium intake before and during training or competition. Consumption of electrolytes in beverages may be prudent during exercise that lasts more than 4 h if sweat sodium and water loss are large. Sweat contains approximately 0.8-2.0 g NaCl/L and 3.0-4.0 g NaCl/L in acclimatized and nonacclimatized individuals, respectively. Second, you should not assume that you can ingest unlimited quantities of fluid during exercise and expect it to be absorbed and distributed to the extracellular and intracellular spaces in a rapid and uniform pattern. To avoid consuming too much or too little fluid, you can simply measure body weight before and after exercise. In fact, some ultraendurance events (e.g., the Western States 100 Mile Race) now require contestants to weigh themselves at regular intervals; competitors are instructed to increase and decrease water consumption accordingly. During prolonged, low-intensity exercise, you usually will lose 0.5-1.5 kg of body water (0.5-1.5 L of sweat) per hour. Therefore, each kilogram of body weight change represents 1 L (and each pound represents a pint) of fluid.

Third, you can estimate the amount of salt (NaCl) you lose during training or competition by using the method shown in table 2.4. Multiply the duration of exercise (hours) by your sweat rate (liters per hour; see figure 2.6 on page 30) to get the total water loss (liters). Then, if you are very fit and well acclimatized to the heat, estimate your sweat salt concentration as 0.8 g NaCl/L; if you are unfit and not acclimatized, estimate this value as 4.0 g NaCl/L of sweat. Next, multiply this sweat salt concentration (grams per liter) by your total water loss to derive the total grams of NaCl lost in sweat. Also, remember that longer events will involve urine NaCl losses. To put this salt loss into perspective, remember that the average adult in America consumes 6-13 g NaCl/day. Clearly, a serious salt deficit is possible in an ultraendurance event.

Table 2.4

Typical Range of Salt (NaCl) and Water Losses of Runners, During Three Common Hot Weather Road Races

Measurement	Fun run (10 k) (6.2 mi)	Marathon (42.2 k) (26.2 mi)	Ultramarathon (161 k) (100 mi)
Duration of exercise (h)	0.5-1	2.1-4.7	20-30
Running pace (min/mi)	5:00-10:00	5:00-10:00	12:00-18:00
Sweat rate (L/h)	1-1.5	1-1.5	0.7-1.2
Total water loss as sweat (L)	0.5-1.5	2-7	14-36
Total salt lost in sweat (g)*	0.4-6.0	1.6-28	11.2-144

*Heredity, diet, and heat acclimatization affect salt levels in sweat. These values assume a range of losses equalling 0.8-4.0 g of NaCl/L sweat.

To select foods and beverages for postexercise dietary replacement, table 2.5 suggests that most commonly used fluids replace very little NaCl, per 8 oz serving. This includes the fluid-electrolyte replacement beverages. Liberal salting of food also provides a simple source of NaCl in the case of large deficits. One teaspoon of table salt equals approximately 8 g of NaCl. Finally, the following recipe may be used to provide sodium and carbohydrate in a large volume for field use: add one packet of unsweetened Kool Aid beverage powder (45 g, 1.6 oz), 2 tablespoons of table salt (NaCl), and 1 kg (2.2 lb) of table sugar to 13.6 L (3 gal) of water.

Table 2.5
Salt (NaCl) Content of Beverages and Liquid Foods

Food items*	NaCl content (g)
Soup, canned	1.9-2.5
Tomato juice	1.5
Fluid-electrolyte replacement beverage	0.2-0.3
1% low-fat milk	0.3
Brewed tea	0.05
Apple juice	0.05
Cranberry juice cocktail	0.03
Cola	0.03
Brewed coffee	<0.03
Orange juice, fresh	<0.03
Grapefuit juice, fresh	<0.03
Beer	<0.03

Note: One U.S. five-cent nickel weighs 5 g.

* 227 ml (8oz) serving.

Other Important Strategies That Optimize Performance

When athletes, laborers, or soldiers develop a plan to counteract heat and humidity, and minimize the physiological effects of hyperthermia and dehydration, the following list of recommendations should be considered. These focus on clothing, environmental conditions, recognizing the warning signals of heat illness, altering training programs, and avoiding factors that reduce exercise-heat tolerance.

1. Your clothing should be lightweight, loose-fitting, and porous, to allow skin cooling via evaporation of sweat and dry heat loss (radiation and convection).

2. Television and radio weather reports provide air temperature and humidity readings. You can use these numbers to determine the approximate risk of heat illness during training or competition each day, using figure 2.9 (page 47). If the intersection of the dry bulb temperature (air temperature) and relative humidity lies in the "moderate risk" zone or higher, athletes should watch

for the warning signals of heat exhaustion or exertional heat-stroke, described in number 3 below. If weather conditions are "high risk" or "very high risk," exercise should be revised as noted in number 4 below.

3. Although the signs and symptoms of heat illness often are unique for each case, because different organs and systems are affected, individuals who work or exercise in a hot environment should be familiar with table 2.6. Anyone who recognizes one or more of these warning signals of heat exhaustion or heatstroke should stop exercise, move to a cool or air-conditioned room, cool the body if necessary, consume fluids, and seek medical assistance. Athletes also should watch for the signs and symptoms of heat illness in training partners, teammates, and fellow competitors. When it is clear that someone else is struggling or losing consciousness, act quickly.

Table 2.6
Warning Signals of Two Forms of Heat Illness

Heat exhaustion	Heatstroke
Headache, irritability	Headache
Tingling or numbness on head, neck, back, limbs	Unconscious, coma*
	Loss of mental clarity*
Chills or shivering	Bizarre behavior*
Great fatigue*	Rapid, strong pulse
Rapid, weak pulse	Hot, red skin*
Pale, moist, cool skin	Profuse sweating in most cases
Dizziness*	
Vomiting, nausea	Fainting
Dehydration*	

* Important; these occur in most cases.

4. When the dual environmental stressors (like heat and humidity; see figure 2.9) exist, reduce your expectations. Reduce the pace, distance, or duration of your workout or race. As noted earlier (see section titled "Heat Acclimatization," page 27), you should confine interval training or speed work to the cooler periods of each day or early morning.

5. After you have acclimatized to the heat, you may experience day-to-day variations in heat tolerance. For example, exercise physiologists have known for years that bacterial or viral illnesses reduce heat tolerance. Other detrimental factors include sleep loss, depletion of carbohydrate (e.g., blood glucose, liver or muscle glycogen), medications (see reference 45 for a complete list), acute or chronic drug/alcohol abuse, and a sudden increase in training or heat exposure.[46] When you recognize that one or more of these factors is present, your expectations and exercise should be scaled back, as in number 4 above. This is not the day to attempt a personal best!

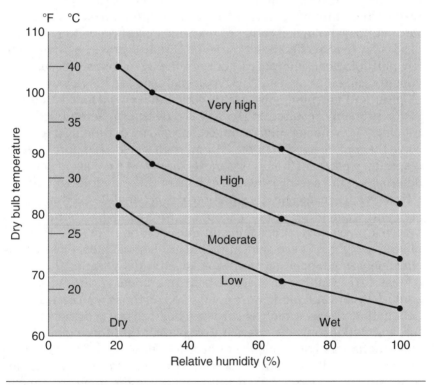

Figure 2.9 Risk of heat exhaustion or heatstroke while racing in hot environments. Reprinted from Armstrong et al. 1996.

Effects of Aging on Heat Tolerance

The investigations that have evaluated adults between the ages of 45 and 64 years suggest that middle-aged men and women are less tolerant of exercise-heat stress, and suffer greater physiological strain during heat acclimation, than younger individuals. Differences such as higher heart rates, higher skin and rectal temperatures, and lower sweat rates were reported in publications appearing between 1956 and 1972.[47] However, none of these studies matched older and younger test subjects for pertinent physiological or morphological factors, such as body weight, surface area, body fat content, and $\dot{V}O_2$max. When middle-aged men and women were matched with younger counterparts for these factors, the resultant heat strain was similar.[48, 49] In fact, Pandolf and colleagues demonstrated that the exercise-heat tolerance of nine younger men (mean age = 21.2 years) was poorer than that of nine middle-aged men (mean age = 46.4 years), in terms of tolerance time (i.e., minutes of exercise at 49°C [120°F], 20% rh to reach a rectal temperature of 40°C [104°F]), final heart rate, sweat rate, and psychological rating of perceived exertion on the first day of a 10-day heat acclimation protocol.[50] By day 8, however, there were few physiological differences between the age groups. Because the younger and older men had been pair-matched for all relevant physiological and morphological traits, the authors concluded that one factor probably resulted in the day 1 differences: aerobic exercise training. The younger men averaged only 4.8 mi of running per week, while the middle-aged men ran 24.1 mi each week. Therefore, cardiorespiratory physical fitness was more important than aging per se in determining the exercise-heat tolerance of these middle-aged men.

However, there may be one situation in which middle-aged and older men and women are more susceptible to greater heat strain. This involves physiologically significant dehydration of greater than 2% of body weight.[47] These same individuals also may not rehydrate as effectively as younger men and women, after dehydration.[51]

Most early studies of exercise-heat tolerance, regarding elderly adults 64 years and older, were conducted on those who had chronic debilitating diseases such as coronary heart disease, diabetes mellitus, and Parkinson's disease.[52] It was evident that these elderly test subjects had a high incidence of heat intolerance, primarily involving underlying disease states, of which the most prominent was cardiovascular disease.[47] Two studies, however, have observed healthy elderly test subjects. The first required that 57 men and 60 women, who ranged in age from 17 to 88 years, walk in the desert for 1 h at an exercise intensity of 40% $\dot{V}O_2$max.[52] The heat-acclimated, fit, healthy, elderly men and women had similar rectal temperature, heart rate, and sweat rate values as younger subjects. The second study evaluated six elderly men (61 to 73 years) and ten young men (21 to 39 years) who rested for up to 130 min in a controlled environment

(40°C [104°F], 40% rh).[53] The healthy, elderly men exhibited no impairment of sweating ability, but apparently had a delayed vasodilation reflex in skin blood vessels, causing greater heat storage. Thus, when the effects of chronic debilitating diseases are kept to a minimum, the heat tolerance of the elderly does not appear to be greatly compromised.[47] The exercise-heat tolerance of elderly adults has not been evaluated during strenuous, high-intensity exercise, however.

MEDICAL CONSIDERATIONS: THE HEAT ILLNESSES

In 1977, the World Health Organization recognized seven different heat illnesses, ranging from swelling (**edema**) of the feet and hands to life-threatening hyperthermia.[54] Four of these illnesses, which occur among athletes and laborers, are described in table 2.3 (page 39): heat exhaustion, exertional heatstroke, heat cramps, and heat syncope. It is interesting that heat acclimatization reduces the incidence of all of these ailments.[14, 22]

Heat Exhaustion

Heat exhaustion is the most common form of heat illness. It is defined as the inability to continue exercise in the heat. The following description of severe heat exhaustion was given by American distance runner Todd Williams, after collapsing in the 5000 m run at the U.S. Olympic Trials in Indianapolis, Indiana, in June 1997:

> I started feeling a little bit of chill and my legs went numb. Then I just went out [fainted]. That's the hardest I've ever run, and the wheels came off.
>
> We were running over our heads for these conditions, but that's what you have to do to improve. I wouldn't want to recommend to everyone that they pass out [faint], but sometimes you have to push the envelope. (Reprinted courtesy of Boston Globe)[55]

The meteorological conditions at the time of this event were 35°C (95°F) and 43% rh ("high risk" category, see figure 2.9). Williams and teammate Bob Kennedy, who won the race in 13:31, were trying to set a pace to finish between 13:10 and 13:15. This was equivalent to running three consecutive miles in 4:14 each, without stopping!

Williams' comment about "running over our heads" is important to your understanding that heat exhaustion involves the inability of the cardiovascular system to sustain strenuous exercise. When challenged with multiple stressors (such as heat, humidity, dehydration, high-intensity exercise), the central nervous system causes adaptive responses in the heart and blood vessels that alter the heart rate and stroke volume (see section above titled "Cardiovascular Responses to Heat and Humidity," page 22). However, when the intensity and duration of exercise are great or when a large dehydration exists, cardiovascular compensation cannot meet the demand of supplying blood to both muscle and skin, and heat exhaustion eventually occurs unless exercise is terminated.

Causes of Heat Exhaustion

For the other 99.9% of humanity, those of us who are not elite athletes, heat exhaustion usually is not due to pushing one's body beyond its limits. It is caused primarily by depletion of circulating fluid (e.g., plasma), which results in inadequate cardiovascular compensation. This explains why it is often said that heat exhaustion is a fluid depletion problem. When fluids are replaced, the medical signs and symptoms vanish.[14, 15] If fluids are replaced orally, electrolyte solutions (e.g., 0.1% NaCl solution) may be used when patients are conscious. In more severe heat exhaustion cases, IV solutions are recommended. A physician will examine pulse, blood pressure, and changes in these measures on standing to determine the contents of the IV solution (i.e., NaCl, dextrose). Rapid recovery occurs after IV fluids are administered for 15-30 min.[14]

Recommendations for Replacing Water and Salt

Two types of heat exhaustion are commonly seen among athletes and soldiers: water depletion (WD) and salt depletion (SD). Although both are caused by a shrunken plasma volume, the most effective medical treatment is slightly different for each. WD occurs rapidly, usually within one day. It involves prominent thirst because blood is concentrated due to excessive water loss in sweat. Dilute fluids or pure water are effective in bringing the body's extracellular fluid back to its normal homeostasis. SD takes 3-5 days to develop because salt losses in sweat and urine eventually exceed the dietary salt content.[56] Thirst is seldom observed in SD; body fluids are dilute and blood sodium

Heat exhaustion is marked by the inability to continue exercise in the heat.

levels are low. However, muscle cramps, vomiting, and progressive weakness are common.[14] To replace the proper nutrients, salt and water losses can be estimated as 2 g of NaCl and 1.5 L of water per hour. The fluids that are used to treat SD should, therefore, contain an appropriate amount of NaCl and should not consist of pure water.

Exertional Heatstroke

Heat exhaustion, characterized by prominent fatigue and progressive weakness, can be distinguished from **exertional heatstroke**, which is a medical emergency involving life-threatening hyperthermia (T_{re} greater than 39-40°C, 103-104°F). Table 2.7 (page 52) compares the distinguishing characteristics of these illnesses. Appendix B offers further information.

Characteristics of Exertional Heatstroke

The first of the distinguishing factors, blood enzymes, are used clinically as markers of organ damage in numerous diseases. Their chemical names are usually abbreviated (e.g., ACT, AST, CPK, LDH). Enzymes are proteins that normally reside inside cells, where they serve to speed chemical reactions. Enzyme levels in the blood rise because damage to cell membranes has occurred. The most likely cause of membrane disruption is hyperthermia—the noxious agent in exertional heatstroke. In fact, if you compare the first two diagnostic factors in table 2.7, you will see that blood enzyme levels rise as T_{re} rises. The third factor in this table relates to the body's ability to spontaneously dissipate heat (see "How Your Body Loses Heat," page 18). In mild cases of heat exhaustion, the rate of heat loss is normal and approximately equals metabolic heat production. In severe cases of heat exhaustion, heat dissipation is overwhelmed by heat production because sweating and cardiovascular compensation do not adequately remove the heat generated during exercise (i.e., when body temperature is moderately elevated). In exertional heatstroke, the central nervous system is *impaired* by the extreme temperature of nerve cells, especially when body temperature is above 41°C (106°F).[15] This scrambles or reduces the number of impulses sent from the brain to the sweat glands and blood vessels, thereby impairing heat loss via sweat evaporation, radiation, and convection. Interestingly, the central nervous system seems to be the most heat sensitive of all of the body's systems. This explains the fourth factor in table 2.7. Mental acuity and consciousness are markedly affected by hyperthermia. This provides the best way to quickly distinguish heat exhaustion from exertional heatstroke. If tests of mental status (e.g., carrying on a coherent conversation, answering simple questions) indicate confusion or loss of acuity, heatstroke should be expected and rapid whole-body cooling should commence immediately. Remember that

Table 2.7
Distinguishing Heat Exhaustion From Exertional Heatstroke

Diagnostic factors	Heat exhaustion		Exertional heatstroke
	Mild	Severe	
Blood enzymes	Normal	Elevated	Elevated
Rectal temperature*	Up to 39-40°C	Variable	Over 39-40°C
Spontaneous body cooling	Present	Variable	Absent
Mental acuity	Normal	Impaired	Disorientation, coma, bizarre bahavior

* Rectal temperature varies with time elapsed between collapse and measurement, cooling efforts, and environmental conditions.

exertional heatstroke remains the greatest threat to *healthy* humans who exercise or work in hot environments.

Recommendations for Cooling Heatstroke Patients and Hyperthermic Athletes

The risk of death or disability by hyperthermia decreases dramatically when cooling is instituted rapidly.[57] Further, the time course of recovery from exertional heatstroke is unique to each patient, due to differences in the duration and magnitude of hyperthermia, as well as the method of cooling employed.[57, 58] For example, figure 2.10 (page 54) depicts two hypothetical athletes who finished strenuous exercise with a severely elevated rectal temperature of 43.3°C (109.9°F). Athlete A (bottom plot) received whole-body cooling therapy because exertional heatstroke was expected. Athlete B (top plot) rested in a hot environment without cooling. Clearly, the "area under the curve" (i.e., the integral) of athlete A represents a much different exposure to heat than that of athlete B.[58]

Cold Water Immersion   For EMTs, physicians, and nurses, the goal of cooling therapy must be to reduce body temperature to a safe level (about 37°C, 98.6°F) as rapidly as possible, in a manner analogous to curve A in figure 2.10. Therefore, it is important to know the cooling power provided by various techniques, as published in the scientific and clinical literature. Figure 2.11 (page 55) provides the cooling rates (°C/min) of seven widely used methods (see figure legend). Obviously, ice-water or cool-water immersion (bars f and g) provides the fastest cooling rate of any known treatment technique.[58] However, this fact has been disputed by several authors since 1959 when Wyndham and colleagues concluded that immersion in cold water (14°C, 58°F) was significantly less effective than other cooling methods. They reported that intense constriction of skin blood vessels created a barrier to heat loss, insulated the body's core, minimized heat loss via conduction and convection, and encouraged heat storage in the deep tissues of the body. However, the temperature difference between very cold water (0.5-4°C, 33-40°F) and skin (typically 34-37°C, 93-96°F) is so great that heat is lost from the body rapidly. This explains why survivors of aircraft and ship crashes (e.g., the *Titanic*) succumb to hypothermia (i.e., core body temperature less than 36°C, 97°F) while floating in the ocean.

Other authorities have expressed concern about ice-water or cool-water immersion because it (a) causes shivering-induced heat production[58] and (b) involves difficulties with specialized treatment such

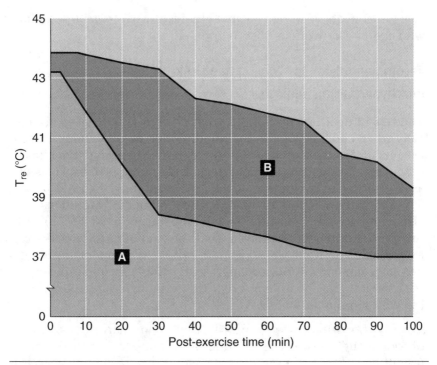

Figure 2.10 Time course of whole-body cooling in two hypothetical hyperthermic athletes.
Reprinted from Armstrong 1999.

as heart defibrillation, unsanitary conditions (i.e., vomiting or diarrhea in the water), and patient discomfort.[59, 60, 61] However, field observations of hyperthermic runners indicated that these issues are minor or nonexistent in the vast majority of exertional heatstroke cases.[62] For example, Professor Carl Maresh, Joyce Armstrong, and I joined the medical directors of the annual Falmouth, Massachusetts, Road Race, Dr. Arthur Crago and Dr. Richard Adams, to study the cooling power of two different therapies. They treat an average of ten exertional heatstroke cases each year. We learned that shivering was not seen in hyperthermic runners, during this midsummer contest, when rectal temperatures were >39°C (102°F) [62] Also, the concerns over difficulties in treatment, which were rare, do not override the primary goal of therapy: to cool the patient as rapidly as possible. Figure 2.11 shows the cooling rate achieved at Falmouth with ice-water immersion (bar g). The cooling provided by resting air exposure while covered with cool, wet towels was similar to bar b in figure 2.11, and required about twice as much time. It is noteworthy that the ice-water modality provided the fastest cooling rate in the scientific and clinical literature, with no instances of complications after treatment, during a 10-year period. This impressive record is surpassed only by Costrini's report of a 0% fatality rate in 252 cases of exertional heatstroke among military recruits.[57] In comparison, the death rate among religious pilgrims, who experienced exertional heatstroke while walking the desert, has been reported at 4.7-18.4%,[63] and up to 80% among elderly patients,[61] when a warm-air spray was utilized (see bar b in figure 2.11).

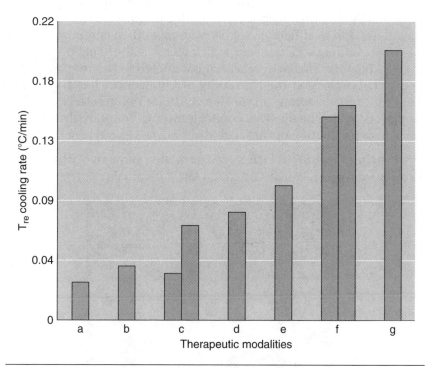

Figure 2.11 Comparison of rectal temperature cooling rates (°C/min) of seven common therapeutic modalities: a, six ice packs at the neck, groin, axillae; b, warm air (45°C) and water (15°C) spray; c, ice packs covering the body; d, 31°C air-water spray; e, 38°C helicopter down-draft; f, 1-3°C ice-water immersion of the body to the sternum; g, partial ice-water immersion (1-3°C) at the Falmouth, Massachusetts, Road Race. Reprinted from Armstrong 1999.

Pouring Water on Skin Because the body is constantly generating and dissipating heat, deep body temperature represents the net turnover of heat from all sources (additive and subtractive). This concept provides the answer to a question that is commonly asked about exercise-induced hyperthermia, "Is spraying or pouring water onto the skin during exercise an effective way to cool the body?" You will note that this question regards cooling *during* exercise, whereas the examples in the two previous paragraphs referred to recovery *after* exercise. The answer is simple. The cooling power of this technique involves evaporation of only the small amount of water that remains on the skin. Compared to the relatively large amount of heat generated in muscles during exercise, spraying or pouring water on the skin has little effect on rectal temperature. This has been demonstrated in two previous laboratory studies that observed runners engaged in treadmill exercise in a hot environment.[64, 65] These studies suggest that temporarily spraying the body, for cooling purposes, is a poor recommendation for an athlete or soldier who must then continue exercise with wet clothing and footwear, which increase the likelihood of chafing and blisters. Similarly, research conducted in the Human Performance Laboratory at the University of Connecticut has shown that nonevaporative cooling, supplied as either a refrigerated headpiece or vest, provided insufficient cooling power to counterbalance metabolic heat production during 30 min of vigorous exercise in the heat.[66]

Preventing Complications A recent review of recovery from exertional heatstroke[58] concluded that a few individuals experience com-

© Paul T. McMahon

Water spray may feel good to over-heated athletes, but it is not the most effective means of lowering their overall body temperatures.

plications, ranging from temporary impairment of single organs or systems[46] to permanent neurological deficits,[67] whereas others—the vast majority—recover completely. This viewpoint is supported by a published classification of patients as either "light" or "severe."[68] The latter group were unconscious or unresponsive on admission and died of central nervous system damage before multiple system disruption became evident. The former group exhibited a brief coma, were cooled within one hour of collapse, but showed blood enzyme elevations nevertheless (table 2.7). The critical factors in this review and virtually all other reports are the speed with which cooling is administered, the cooling power of the modality employed, and the "area under the heating curve" (figure 2.10). This explains why none of the 100 runners at the Falmouth Road Race (10-year period)[62] and none of the 252 recruits at the Parris Island, South Carolina, Marine base (10-year period)[57] died after experiencing exertional heatstroke with rectal temperatures that ranged up to 42.8°C (109°F). They were cooled rapidly via ice-water immersion, and the area under their heating curves decreased before multisystem tissue injury occurred. This outstanding treatment record exemplifies an impressive history of improved education and training, revision of doctrines, and updated medical care. Since 1956, an extensive program for prevention of heat casualties among recruits has been in operation at Parris Island.[69] But this was not always true. During World War II, when heatstroke was not as widely recognized and preventive measures were not as refined, nearly 200 fatal cases of exertional heatstroke occurred in military (all branches) training centers within the United States. The most susceptible individuals were overweight recruits from homes in the northern United States, undergoing their first weeks of summer training at camps located in the southern states.[70] Another classic, but sad, example of the potential of heatstroke to cause casualties and deaths is illustrated in table 2.8. These statistics were derived from World War II medical records of the armed forces of the United Kingdom,[71] regarding a time when few preventive measures were taken against heatstroke. During these seven years, the incidence was approximately 1 case per 1,000 soldiers. This rate is very similar to the incidence of exertional heatstroke observed at mass participation footraces on roads.[72]

Heat Cramps

During prolonged or repetitive exercise, athletes and laborers may occasionally experience **heat cramps**. These excruciating contractions

Table 2.8
Number of Heatstroke Cases in the Armed Forces of the United Kingdom, from 1939-1945, and Their Outcomes

Year	Heatstroke cases	Deaths	Subsequent days of sickness	Heatstroke incidence (per 1000 soldiers)
1939	121	2	601	0.9
1940	218	1	1095	0.8
1941	529	2	1860	1.3
1942	344	8	2881	0.7
1943	314	1	1773	0.5
1944	575	6	3267	0.7
1945	854	14	4728	1.1
Totals	2955	34	16,205	mean = 0.9

Adapted from Mellor 1972.

of skeletal muscle are linked with either large losses of NaCl in sweat, replacement of sweat losses with a large volume of dilute fluid/pure water, or both. All of these situations cause the extracellular fluid, and the intracellular fluid after a few hours, to become diluted. The expansion of intramuscular water is believed to be the critical factor in the development of heat cramps.[73, 74] Indeed, an increase of intramuscular water has been observed during heat acclimation while consuming a low-sodium diet, via analysis of muscle biopsies.[24]

A shift in the intracellular-extracellular ratio of NaCl and water will alter the electrical properties of muscle membranes, and could alter muscle contraction or relaxation. Although the exact mechanism of any muscle cramp (e.g., writer's cramp, nocturnal foot cramp) is unknown, research suggests that heat cramps are related to salt imbalance. For example, Leithead and Gunn[75] reported that the urine of healthy field laborers contained 36 g NaCl/L, but the urine of laborers with heat cramps contained only 4 g/L. The smaller amount of NaCl in the urine of the heat cramp patients suggests that their kidneys were reabsorbing sodium, under the influence of the hormone aldosterone, because of the whole-body deficit. Similarly, Talbot[76] observed blood sodium levels ranging from 121 to 140 mEq/L in 32 heat cramp patients; because the normal range in healthy adults is 135 to 145 mEq/L, these patients showed severe to moderate sodium deficits.

Heat cramps may be confused with other forms of muscle cramps.[14] They usually occur in the large muscles, limbs, or in the abdominal muscles. Unlike ordinary exercise-induced muscle cramps, heat cramps begin as feeble, localized spasms and may appear to wander over the muscle, as adjacent motor units become involved.[73] They occur in 50-70% of SD heat exhaustion patients.[14]

Acclimatization and Heat Cramps

Heat acclimatization has been reported to have no effect on the development of heat cramps,[77] and it has been suggested that they occur more often in acclimatized, highly trained athletes.[78] However, these reports cite no verifying evidence and they do not agree with the following observations made in field and industrial settings:[14]

• Very few cases of heat cramps were observed among soldiers during many years of heat illness triage in Israel (0% incidence)[68] and India (1.5% incidence).[79]

• Twenty-six laboratory tests on one subject (i.e., 2 h of exercise-heat exposure during negative salt balance) supported improved resistance to heat cramps after heat acclimatization.[73]

• The incidence of heat cramps was greatest during the *first few days* of a heat wave at the Boulder Dam Construction site in Nevada, in the early 1930s.[80]

• Steel workers, who work in hot indoor factories, were maximally susceptible to heat cramps during the *first few days* of an outdoor heat wave that occurred in midsummer in Youngstown, Ohio.[76]

• A tennis player, who had been suffering from painful heat cramps for months, found permanent relief after consuming more NaCl in his food and fluid-electrolyte beverages.[81]

These observations support the concept that heat acclimatization enhanced physical fitness, and the body's hormone-mediated conservation of sodium all serve to reduce the incidence of heat cramps.

Causes of Heat Cramps

Because heat cramps cannot be induced in humans today, due to ethical considerations involving pain and illness in human test subjects, controlled studies that focus on the cause(s) and ultrastructural mechanism(s) may never be conducted. However, several promising areas for future research have been proposed,[14] based on previous anecdotal reports, clinical retrospective studies, and case reports. For example, the work of Ladell showed that pain and cramping was eliminated in one leg, but not the other, when a small volume of concentrated saline solution (15% NaCl) was given intravenously.[73] The former leg had no treatment applied, but the latter had blood flow occluded by a pneumatic tourniquet (250 mm mercury). This study suggested that heat cramps resulted from a disturbance of fluid-electrolyte balance at the muscle, not in inner organs. Further, recognized causes of other forms of muscle cramps might be investigated to clarify the mechanism(s) of heat cramps. These other causes and their associated illnesses include disorders of nerve axons (e.g., diabetes, alcohol abuse), metabolic defects (e.g., McArdle's syndrome, enzyme deficiencies), and cold-air or cold-water exposure (e.g., swimmer's cramps).[82, 83]

Recommendations for Replacing Fluids and Electrolytes

The treatment of heat cramps involves replacing NaCl and water to restore the homeostatic equilibrium on both sides of the muscle membrane, as described in table 2.3 (page 39). The use of IV solutions is rapid and effective.[74, 84] Oral NaCl replacement involves consuming a dilute salt solution containing two salt tablets (weighing 0.6 g or 10 grains each), crushed and dissolved in 1 L of water. Thus, it is not surprising that steel mill managers have reported successfully reducing the incidence of heat cramps to nearly zero by encouraging table salt consumption.[76] Nor is it surprising that laborers in Germany and England salt their beverages to successfully prevent heat cramps.[74]

Heat Syncope (Fainting)

The final heat illness in this chapter, **heat syncope**, does not occur among those who are physically fit as often as it does among unfit, sedentary, nonacclimatized individuals. In fact, the two most common times for heat syncope to occur are (a) when laborers, soldiers, civilians, or band members stand erect for a long period of time, and (b) after ending strenuous, prolonged exercise in an upright posture. This stems from the fact that syncope is an illness caused by insufficient blood flow to the brain. In both of these instances (a and b above), standing in a hot environment results in pooling of blood in the veins of the limbs and skin (i.e., increased skin temperature causes maximal cutaneous vessel dilation), leading to an insufficient volume of blood returning to the heart. Because cardiac filling decreases, the cardiac output and blood pressure both decline, resulting in insufficient oxygen delivery to the brain.[14]

The medical diagnosis of heat syncope is based on a brief fainting episode in the absence of an elevated rectal temperature. The pulse and breathing rates are slow, and skin pallor is obvious. Before the syncopal episode, a patient may experience sensations of weakness, tunnel vision, vertigo, or nausea. After syncope occurs, a horizontal posture encourages venous blood to return to the heart, causing an increase in cardiac output, blood pressure, and delivery of blood to the brain. Medical treatment and first-aid measures include lying with feet elevated, replacing fluid and electrolyte deficits, and avoiding sudden or prolonged standing.

Syncope and Acclimatization

The incidence of heat syncope declines with each day of exercise-heat exposure.[85] This parallels the circulatory adaptations that occur during the first 3-5 days of heat acclimatization (i.e., plasma volume expansion, decreased heart rate; see table 2.2). This phenomenon is illustrated in figure 2.12, which has been redrawn from a very early study of the body's adaptive responses during heat acclimation.

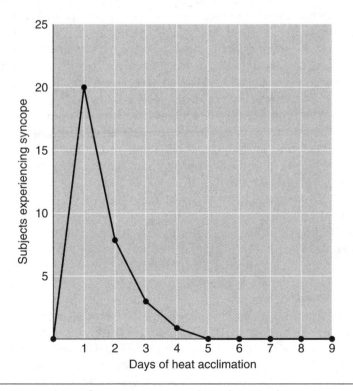

Figure 2.12 Incidence of syncope among 45 subjects living in a hot environment for 24 h each day and undergoing exercise trials.
Reprinted from Bean and Eichna 1943.

Forty-five male test subjects lived in the heat for 9 days, and underwent lengthy exercise trials each day. After nearly half of these males experienced syncope on day 1, investigators saw a rapid decline in these events until days 5-9, when no one fainted. Although heat syncope may be associated with heat exhaustion and sometimes may be treated by replacing lost water and salt, heat syncope usually is categorized as a distinct syndrome because fainting may occur in individuals who have normal fluid-electrolyte status.[74, 83, 84]

Causes of Heat Syncope

If fainting occurs after exercise, when environmental conditions are mild, heat syncope per se is not involved. The possible causes for this type of syncope are numerous, and are classified as either cardiac or noncardiac. The former category includes heart valve obstruction, electrical conduction abnormalities (e.g., arrhythmias), and nerve dysfunction. The latter includes metabolic (e.g., low blood sugar), neurologic (e.g., seizure), and psychiatric causes.[86] However, *vasovagal syncope* is probably the most common type of fainting seen among athletes. The term "vasovagal" refers to the effect of the vagus nerves on veins and the heart. These nerves emerge from the medulla and pass downward through the neck, near the jugular vein, to various autonomic ganglia. From there, the vagus fibers innervate muscles of the thoracic and abdominal organs, where they control heart rate and almost all abdominal activities.[87] Vigorous contractions of the ventricle in response to prolonged standing, dehydration, exercise, and increased blood levels of the hormone epinephrine (see chapter 1) are believed to initiate a vasovagal reaction by activating cardiovascular receptors that sense movement and/or force. These receptors are thought to initiate a reflex in the brain (specifically, the medulla) that produces vasovagal syncope. This state results from a strong activation of parasympathetic signals to slow the heart and sympathetic signals to dilate peripheral blood vessels, causing decreased cardiac output and blood pressure plus the classic symptoms of pallor, nausea, blurred vision, and fainting.[88] The enhanced tone of the vagus nerve in endurance athletes probably explains why they are more susceptible to exercise-induced vasovagal syncope than nonathletes. Although this type of fainting is not life-threatening, the falls resulting from such vasovagal events can be dangerous. When athletes experience syncope repeatedly, it is more likely that cardiac complications or disease are involved, but proper diagnosis and medications may improve the possibility that the athletes can return to their sport.[88]

Purposeful Athletic Weight Loss and Dehydration

Some individual sports are contested in a spectrum of weight classes. This prompts some athletes to lose weight by restricting food and fluid intake, and by resting or exercising in a hot environment. A small number of weight lifters, rowers, boxers, and wrestlers have used diuretics (water pills that reduce reabsorption of water in the kidneys) to speed weight loss by increasing urine production. The International Olympic Committee and the National Collegiate Athletic Association have banned the use of diuretics. Despite these regulations, 41% of collegiate wrestlers have reported large weight fluctuations (5-9 kg) every week of their competitive season.[89] This was painfully demonstrated by the deaths of three American collegiate wrestlers during fall of 1997. Diuretic use results in great reductions in cardiovascular endurance[90] and larger fluid loss from the circulatory system than any other form of dehydration.[14, 19] When only food restriction is used, water and lean body mass are the primary substances to be lost, not fat. The result is a decrease of muscle carbohydrate stores (e.g., glycogen) and muscle water, which translates to decrements of temperature regulation, cardiovascular function, endurance, and possibly power when near-maximal exercise lasts longer than 30 sec.[7] Other methods of rapid weight loss appear to affect athletic performance in different ways. For example, severe fluid restriction may alter the performance of strength athletes. When exercise in a hot environment is combined with water restriction, decreases in strength and power are even more likely to result.[20]

Even psychological factors associated with rapid weight loss may influence physical performance negatively.[21] The resultant mood changes and altered decision making[20] suggest that athletes who dehydrate to "make weight" could be at a psychological disadvantage when compared to opponents who do not undergo dehydration. These effects may be partly reduced by undergoing rapid weight loss many times.[21]

 ## *Recommendations for Counteracting Heat and Humidity*

The following questions have been developed as a summary of this chapter.[7,22,29,58,72] Athletes, soldiers, and laborers alike can use this checklist to evaluate their readiness to perform exercise, work, or live in a hot

environment. Corrective action should be taken if any question is answered "no."

- Have I trained for 10-14 days in a hot environment, similar in temperature and humidity to the ultimate target environment, before I begin competition or an 8 h work shift?

- Do I wear lightweight, loose-fitting clothing?

- Do I know the signs and symptoms of heat exhaustion, exertional heatstroke, heat syncope, and heat cramps?

- Do I avoid lengthy warm-up periods on very hot-humid days? (for athletes only)

- Do I know my sweat rate and the amount of fluid that I should drink to replace every kilogram of body weight lost? Do I drink an appropriate volume of fluid before, during, and after exercise?

- When training requires that I cover many miles, do I (a) design a course that allows frequent rehydration at water fountains, schools, gas stations, convenience stores, and city parks, or (b) carry water in bottles held by a belt, pouch, holster, or backpack?

- Is my body weight today within 1% of my average body weight?

- Do I know how to monitor my body's core temperature?

- Is my urine output plentiful and is the color "very pale yellow," "pale yellow," or "straw colored"?

- Is my dietary intake of salt adequate?

- Do I perform intense exercise (e.g., interval training) during the cooler hours of the day or in the early morning?

- Do I avoid strenuous exercise in the heat after experiencing those factors that lower exercise-heat tolerance (e.g., sleep loss, infectious illness, fever, diarrhea, vomiting, carbohydrate depletion, some medications, alcohol or drug abuse)?

- Have I checked the weather forecast and used the temperature and humidity to estimate my risk of heat illness? Do I reduce my training when the risk is moderate, high, or very high?

- Do I know where fluids and first aid are available at the race course, training site, or worksite?

- Have I developed and tested a plan to avoid dehydration and the ill effects of exercise/work in the heat?

REFERENCES

1. Sawka, M.N.; Wenger, C.B.; & Pandolf, K.B. 1996. Thermoregulatory responses to acute exercise-heat stress and heat acclimation. In *Handbook of physiology*. Sect. 4: *Environmental physiology*. Vol. 1, edited by M.J. Fregly & C.M. Blatteis, 157-185. New York: Oxford University Press.

2. Jessen, C. 1996. Interaction of body temperatures in control of thermoregulatory effector mechanisms. In *Handbook of physiology*. Sect. 4: *Environmental physiology*. Vol. 1, edited by M.J. Fregly & C.M. Blatteis, 127-138. New York: Oxford University Press.

3. Haymes, E.M., & Wells, C.L. 1986. *Environment and human performance*. Champaign, IL: Human Kinetics, 1-164.

4. Adams, W.C.; Fox, R.H.; Fry, A.J.; & MacDonald, I.C. 1975. Thermoregulation during marathon running in cool, moderate and hot environments. *Journal of Applied Physiology 38*: 1030-1037.

5. Armstrong, L.E. 1991. Body temperature regulation. *National Strength and Conditioning Association Journal 13*: 66-67.

6. Nadel, E.R. 1977. A brief overview. In *Problems with temperature regulation during exercise*, edited by E.R. Nadel, 1-26. New York: Academic Press.

7. Armstrong, L.E. 1992. *Keeping your cool in Barcelona, a detailed report*. Colorado Springs: U.S. Olympic Committee, Sport Sciences Division, 1-29.

8. Rowell, L.B. 1993. *Human cardiovascular control*. New York: Oxford University Press.

9. Wyndham, C.H., & Strydom, N.B. 1969. The danger of inadequate water intake during marathon running. *South African Medical Journal 43*: 893-896.

10. Davies, C., & Young, K. 1983. Effect of temperature on muscle energy metabolism and endurance during successive isometric contractions, sustained to fatigue, of the quadriceps muscle in man. *Journal of Physiology 220*: 335-352.

11. Segal, S.S.; Faulkner, J.A.; & White, T.P. 1986. Skeletal muscle fatigue *in vitro* is temperature dependent. *Journal of Applied Physiology 61*: 660-665.

12. Sargeant, A.J. 1987. Effect of muscle temperature on leg extension force and short-term power output in humans. *European Journal of Applied Physiology 56*: 693-698.

13. Sawka, M.N., & Wenger, C.B. 1988. Physiological responses to acute exercise-heat stress. In *Human performance physiology and environmental medicine at terrestrial extremes*, edited by K.B. Pandolf, M.N. Sawka, & R.R. Gonzalez, 97-151. Indianapolis: Benchmark Press.

14. Hubbard, R.W., & Armstrong, L.E. 1988. The heat illnesses: Biochemical, ultrastructural, and fluid-electrolyte considerations. In *Human performance physiology and environmental medicine at terrestrial extremes*, edited by K.B. Pandolf, M.N. Sawka, & R.R. Gonzalez, 305-360. Indianapolis: Benchmark Press.

15. Hubbard, R.W., & Armstrong, L.E. 1989. Hyperthermia: New thoughts on an old problem. *The Physician and Sportsmedicine 17*: 97-113.

16. Rowell, L.B. 1986. *Human circulation regulation during physical stress*. New York: Oxford University Press.

17. Sawka, M.N. 1988. Body fluid responses and hypohydration during exercise-heat stress. In *Human performance physiology and environmental medicine at terrestrial extremes*, edited by K.B. Pandolf, M.N. Sawka, & R.R. Gonzalez, 227-266. Indianapolis: Benchmark Press.

18. Armstrong, L.E.; Hubbard, R.W.; Jones, B.H.; & Daniels, J.T. 1986. Preparing Alberto Salazar for the heat of the 1984 Olympic marathon. *The Physician and Sportsmedicine 14*: 73-81.

19. Coyle, E.F., & Hamilton, M.A. 1990. Fluid replacement during exercise: Effects on physiological homeostasis and performance. In *Fluid homeostasis during exercise. Perspectives in exercise science and sports medicine*. Vol. 3, edited by C.V. Gisolfi & D.R. Lamb, 281-308. Carmel, IN: Benchmark Press.

20. Sawka, M.N., & Pandolf, K.B. 1990. Effects of body water loss on physiological function and exercise performance. In *Fluid homeostasis during exercise. Perspectives in exercise science and sports medicine*. Vol. 3, edited by C.V. Gisolfi & D.R. Lamb, 1-38. Carmel, IN: Benchmark Press.

21. Horswill, C. 1991. Does rapid weight loss by dehydration adversely affect high-power performance? *Gatorade Sports Science Exchange 4*: 1-4.

22. Armstrong, L.E., & Maresh, C.M. 1991. The induction and decay of heat acclimatization in trained athletes. *Sports Medicine (New Zealand) 12*: 302-312.

23. Gonzalez, R.R.; Pandolf, K.B.; & Gagge, A.P. 1974. Heat acclimation and decline in sweating during humidity transients. *Journal of Applied Physiology 36*: 419-425.

24. Armstrong, L.E.; Costill, D.L.; & Fink, W.J. 1987. Changes in body water and electrolytes during heat acclimation: Effects of dietary sodium. *Aviation, Space and Environmental Medicine 58*: 143-148.

25. Armstrong, L.E., & Dziados, J.E. 1986. Effects of heat exposure on the exercising adult. In *Sports Physical Therapy*, edited by D.B. Bernhardt, 197-214. New York: Churchill Livingstone.

26. Sciaraffa, D.; Fox, S.C.; Stockman, R.; & Greenleaf, J.E. 1981. Human acclimation and acclimatization to heat: A compendium of research, 1968-1978 (Technical Report No. 81181). Moffett Field, CA: National Aeronautics Space Administration.

27. Francesconi, R.P.; Armstrong, L.E.; Leva, N.; Moore, R.; Szylk, P.C.; Matthew, W.T.; Curtis, W.C.; Hubbard, R.W.; & Askew, E.W. 1991. Endocrinological responses to dietary salt restriction during heat acclimation. In *Nutritional needs in hot environments*, edited by B.M. Marriott, 259-275. Washington, DC: National Academy Press.

28. Wenger, C.B. 1988. Human heat acclimatization. In *Human performance physiology and environmental medicine at terrestrial extremes*, edited by K.B.

Pandolf, M.N. Sawka, & R.R. Gonzalez, 153-198. Indianapolis: Benchmark Press.

29. American College of Sports Medicine. 1996. Position stand. Exercise and fluid replacement. *Medicine & Science in Sports & Exercise 28(1)*: i-vii.

30. Armstrong, L.E.; Maresh, C.M.; Gabaree, C.V.; Hoffman, J.R.; Kavouras, S.A.; Kenefick, R.W.; Castellani, J.W.; & Ahlquist, L.E. 1997. Thermal and circulatory responses to exercise: Effects of hypohydration, dehydration, and water intake. *Journal of Applied Physiology 82*: 2028-2035.

31. Armstrong, L.E., & Maresh, C.M. 1996. Fluid replacement during exercise and recovery from exercise. In *Body fluid balance. Exercise and sport*, edited by E.R. Buskirk & S.M. Puhl, 259-281. Boca Raton, FL: CRC Press.

32. Armstrong, L.E. 1994. Considerations for replacement beverages: Fluid-electrolyte balance and heat illness. In *Fluid replacement and heat stress*, edited by B.M. Marriott, 37-54. Washington, DC: National Academy Press, Food and Nutrition Board, Institute of Medicine.

33. Armstrong, L.E.; Maresh, C.M.; Castellani, J.W.; Bergeron, M.F.; Kenefick, R.W.; La Gasse, K.E.; & Riebe, D. 1994. Urinary indices of hydration status. *International Journal of Sport Nutrition 4*: 265-279.

34. Armstrong, L.E.; Herrera-Soto, J.A.; Hacker, F.T.; Casa, D.J.; Kavouras, S.A.; & Maresh, C.M. 1998. Urinary indices during dehydration, exercise and rehydration. *International Journal of Sport Nutrition 8*: 345-355.

35. Ross, D.L., & Neely, A.E. 1983. *Textbook of urinalysis and body fluids*. Norwalk, CT: Appleton-Century-Crofts.

36. Mickelsen, O. 1982. The nutritional importance of sodium. In *Sodium intake-dietary concerns*, edited by T.M. Freeman & O.W. Gregg, 1-70. St. Paul: American Association of Cereal Chemists.

37. National Research Council. 1989. *Recommended dietary allowances.* 10th ed. Washington, DC: National Academy Press, 250-255.

38. Dahl, L.K. 1972. Salt and hypertension. *American Journal of Clinical Nutrition 25*: 234-244.

39. Hubbard, R.W.; Armstrong, L.E.; Evans, P.K., & De Luca, J.P. 1986. Long-term water and salt deficits: A military perspective. In *Predicting decrements in military performance due to inadequate nutrition*, 29-46. Washington, DC: National Academy Press.

40. Armstrong, L.E.; Costill, D.L.; Fink, W.J.; Bassett, D.; Hargreaves, M.; Nishibata, I.; & King, D.S. 1985. Effects of dietary sodium on body and muscle potassium content during heat acclimation. *European Journal of Applied Physiology 54*: 391-397.

41. Armstrong, L.E.; Hubbard, R.W.; Askew, E.W.; De Luca, J.P.; O'Brien, C.; Pasqualicchio, A.; & Francesconi, R.P. 1993. Responses to moderate and low sodium diets during exercise-heat acclimation. *International Journal of Sport Nutrition 3*: 207-221.

42. Conn, J.W. 1949. Acclimatization to heat. *Advances in Internal Medicine 3*: 373-393.

43. Armstrong, L.E.; Curtis, W.C.; Hubbard, R.W.; Francesconi, R.P.; Moore, R.; & Askew, E.W. 1992. Symptomatic hyponatremia during prolonged exercise in heat. *Medicine & Science in Sports & Exercise 25*: 543-549.

44. Noakes, T.D. 1992. Hyponatremia during distance running: A physiological and clinical interpretation. *Medicine & Science in Sports & Exercise 24*: 403-405.

45. Epstein, Y. 1990. Heat intolerance: Predisposing factor or residual injury? *Medicine & Science in Sports & Exercise 22*: 29-35.

46. Armstrong, L.E.; De Luca, J.P.; & Hubbard, R.W. 1990. Time course of recovery and heat acclimation ability of prior exertional heatstroke patients. *Medicine & Science in Sports & Exercise 22*: 36-48.

47. Pandolf, K.B. 1997. Aging and human heat tolerance. *Experimental Aging Research 23*: 69-105.

48. Armstrong, C.G., & Kenney, W.L. 1993. Effects of age and acclimation on responses to passive heat exposure. *Journal of Applied Physiology 75*: 2162-2167.

49. Kenney, W.L. 1988. Control of heat-induced cutaneous vasodilation in relation to age. *European Journal of Applied Physiology 57*: 120-125.

50. Pandolf, K.B.; Cadarette, B.S.; Sawka, M.N.; Young, A.J.; Francesconi, R.P.; & Gonzalez, R.R. 1988. Thermoregulatory responses of middle-aged and young men during dry-heat acclimation. *Journal of Applied Physiology 65*: 65-71.

51. Mack, G.W.; Weseman, C.A.; Langhans, G.W.; Scherzer, H.; Gillen, C.M.; & Nadel, E.R. 1994. Body fluid balance in dehydrated healthy older men: Thirst and renal osmoregulation. *Journal of Applied Physiology 76*: 1615-1623.

52. Levine, J.A. 1969. Heat stroke in the aged. *American Journal of Medicine 47*: 251-258.

53. Sagawa, S.; Shiraki, K.; Yousef, M.K.; & Miki, K. 1988. Sweating and cardiovascular responses of aged men to heat exposure. *Journal of Gerontology 43*: M1-M8.

54. World Health Organization. 1977. *Manual of the international statistical classification of diseases, injuries, and causes of death.* Geneva, Switzerland: World Health Organization, 535.

55. Huebner, B. 1997. Back on his feet. *Boston Globe*, 16 June, 2(D).

56. Armstrong, L.E.; Hubbard, R.W.; Szlyk, P.C.; Sils, I.V.; & Kraemer, W.J. 1988. Heat intolerance, heat exhaustion monitored: A case report. *Aviation, Space, and Environmental Medicine 59*: 262-266.

57. Costrini, A. 1990. Emergency treatment of exertional heatstroke and comparison of whole body cooling techniques. *Medicine & Science in Sports & Exercise 22*: 15-18.

58. Armstrong, L.E. 1998. Can humans avoid and recover from exertional heatstroke? In *Adaptation biology and medicine.* Vol. 2, edited by K.B. Pandolf, N. Takeda, & P.K. Singal. Japan: Narosa Publishing House, in press.

59. Kielblock, A.J.; Van Rensburg, J.P.; & Franz, R.M. 1986. Body cooling as a method for reducing hyperthermia. An evaluation of techniques. *South African Medical Journal 69*: 378-380.

60. Wyndham, C.H.; Strydom, N.B.; Cooks, H.M.; Maritz, J.S.; Morrison, J.F.; Fleming, P.W.; & Ward, J.S. 1959. Methods of cooling subjects with hyperpyrexia. *Journal of Applied Physiology 14*: 771-776.

61. Khogali, M.; El-Sayed, H.; Amar, M.; Sayed, S.E.; Habaski, S.A.; & Mutwali, A. 1983. Management and therapy regimen during cooling in the recovery room at different heatstroke treatment centers. In *Heatstroke and temperature regulation*, edited by M. Khogali & J.R.S. Hales, 149-156. New York: Academic Press.

62. Armstrong, L.E.; Crago, A.E.; Adams, R.; Roberts, W.O.; & Maresh, C.M. 1996. Whole-body cooling of hyperthermic runners: Comparison of two field therapies. *American Journal of Emergency Medicine 14*: 355-358.

63. Khogali, M., & Al-Marzoogi, A. 1983. Prevention of heatstroke: Is it plausible? In *Heatstroke and temperature regulation*, edited by M. Khogali & J.R.S. Hales, 293-304. New York: Academic Press.

64. Gisolfi, C.V., & Copping, J.R. 1974. Thermal effects of prolonged treadmill exercise in the heat. *Medicine & Science in Sports & Exercise 6*: 108-113.

65. Bassett, D.R.; Nagle, F.J.; Mokerjee, S.; Dan, K.C.; Ng, A.V.; Voss, S.G.; & Napp, J.P. 1987. Thermoregulatory responses to skin wetting during prolonged treadmill running. *Medicine & Science in Sports & Exercise 19*: 28-32.

66. Armstrong, L.E.; Maresh, C.M.; Riebe, D.; Kenefick, R.W.; Castellani, J.W.; Senk, J.M.; Echegaray, M.; & Foley, M.F. 1995. Local cooling in wheelchair athletes during exercise-heat stress. *Medicine & Science in Sports & Exercise 27*: 211-217.

67. Mehta, A.C., & Baker, R.N. 1970. Persistent neurological deficits in heat stroke. *Neurology 20*: 336-340.

68. Shibolet, S.; Lancaster, M.C.; & Danon, Y. 1976. Heatstroke: A review. *Aviation, Space, and Environmental Medicine 47*: 280-301.

69. Minard, D. 1967. Studies and recent advances in military problems of heat acclimatization. *Military Medicine 132*: 306-315.

70. Schickele, E. 1947. Environment and fatal heatstroke. An analysis of 157 cases occurring in the Army of the U.S. during WWII. *Military Surgeon 100*: 235-256.

71. Mellor, W.F. 1972. *Casualties and medical statistics*. London: Her Majesty's Stationary Office, 20-51.

72. American College of Sports Medicine. 1996. Position stand. Heat and cold illnesses during distance running. *Medicine & Science in Sports & Exercise 28(12)*: i-x.

73. Ladell, W.S.S. 1949. Heat cramps. *Lancet 2*: 836-839.

74. Leithead, C.S., & Lind, A.R. 1964. *Heat stress and heat disorders*. Philadelphia: Davies.

75. Leithead, C.S., & Gunn, E.R. 1964. The aetiology of cane cutters cramps in British Guiana. In *Environmental physiology and psychology in arid conditions*, 13-17. Leige, Belgium: UNESCO.

76. Talbot, J.H. 1935. Heat cramps. In *Medicine*, 323-376. Baltimore: Williams & Wilkins.

77. Knochel, J.P., & Reed, G. 1987. Disorders of heat regulation. In *Clinical disorders, fluid and electrolyte metabolism*, edited by C.R. Kleeman, M.H. Maxwell, & R.G. Narin, 1197-1232. New York: McGraw-Hill.

78. Sandor, R.P. 1997. Heat illness. On-site diagnosis and cooling. *The Physician and Sportsmedicine 25*: 35-40.

79. Malhotra, M.S., & Venkataswamy, Y. 1974. Heat casualties in the Indian Armed Forces. *Indian Journal of Medical Research 62*: 1293-1302.

80. Talbot, J.H., & Michelsen, J. 1933. Heat cramps. *Journal of Clinical Investigation 12*: 533-535.

81. Bergeron, M.F. 1996. Heat cramps during tennis: A case report. *International Journal of Sport Nutrition 6*: 62-68.

82. Strauss, R.H. 1984. Skeletal muscle abnormalities associated with sports activities. In *Sports medicine*, 159-164. Philadelphia: Saunders.

83. Dinman, B.D., & Horvath, S.M. 1984. Heat disorders in industry: A reevaluation of diagnostic criteria. *Journal of Occupational Medicine 26*: 489-495.

84. Callaham, M.L. 1979. Emergency management of heat illness. In *Emergency Physician Series*, 1-23. North Chicago, IL: Abbott Laboratories.

85. Bean, W.B., & Eichna, L.W. 1943. Performance in relation to environmental temperature. *Federation Proceedings 2*: 144-158.

86. Benditt, D.G.; Sakaguchi, S.; & Shultz, J.J. 1993. Syncope: Diagnostic considerations and role of tilt table testing. *Cardiology Reviews 1*: 146-156.

87. Guyton, A.C., & Hall, J.E. 1996. *Textbook of medical physiology*. 9th ed. Philadelphia: Saunders.

88. Wang, D.; Sakaguchi, S.; & Babcock, M. 1997. Exercise-induced vasovagal syncope. Limiting the risks. *The Physician and Sportsmedicine 25*: 64-74.

89. Rosen, L.W.; McKeag, D.B.; Hough, D.O.; & Curley, V. 1988. Pathogenic weight-control behavior in female athletes. *The Physician and Sportsmedicine 14*: 79-86.

90. Armstrong, L.E.; Costill, D.L.; & Fink, W.J. 1985. Influence of diuretic-induced dehydration on competitive running performance. *Medicine & Science in Sports & Exercise 17*: 456-461.

Cold, Windchill, and Water Immersion

Temperatures ranged from −47.2°F [−44°C] in the morning to −38°F [−39°C] in the evening. We [walked] 6 3/4 miles in the day.... We have dim daylight from about 10:30 A.M., when dawn begins, to about 2:30 or 3:00 P.M. ... After lunch we march in the dark. Our hands give us more pain with the cold than any other part—our feet are generally warm, but our hands are often dreadfully cold all night.... [In] wet mits the skin is sodden ... and in this condition they get very easily frozen.... [My companion's] finger tips are all pretty badly blistered. I have only one bad thumb which blistered early and is now broken and very sore.

—Edward Wilson, 1911, British Antarctic Expedition[1]

The American vessels were not equipped to ward off the stinging cold, which grew fiercer as the winter advanced. Food congealed. Barrels of fruit had to be chopped apart with an axe. Sauerkraut resembled mica or slate. Butter and lard had to be carved with a cold chisel and mallet. When one seaman tried to bite into an icicle, a piece of it froze to his tongue. Two others lost all the skin on their lips. Facial hair turned to cardboard, and if a man stuck out his tongue it froze to his beard; contact with the metal of a gun penetrated two layers of wool mittens with a sensation of scalding water.

It was dangerous to walk too far from the ship over the rumpled ice. The frost, Kane found, seemed to extend to the brain. An inertia crept over the system; the desire to sit down and rest was almost uncontrollable; drowsiness and death could easily follow.

—Elisha Kane, 1850, British Expedition of the Arctic Channels[2]

Your body responds to cold in unique, necessary ways. Retention of metabolic heat is the primary goal. Two physiological responses seek to restore thermal homeostasis: reduced heat loss via constriction of skin blood vessels, and increased metabolic heat production via shivering. Interestingly, these responses and the effects of cold on exercise performance differ slightly in cold air versus cold-water immersion. The process of cold acclimatization further enhances the homeostatic protection of your body. This chapter also describes tissue injuries that occur during work and athletic endeavors in cold environments. Most of these can be avoided.

COLD STRESSORS

The accounts that open this chapter were taken from the diaries of two courageous explorers who were among the first Europeans to explore the vast wastelands of the Arctic and Antarctic. Because these explorers traveled 87 and 148 years ago, respectively, it is fair to ask what relevance their experiences have for athletes, soldiers, and laborers today. The answer is "great relevance." Even with the development of modern fabrics and insulated gear, intense or prolonged cold exposure can still disrupt the maintenance of body temperature homeostasis, impair pulmonary function via bronchospasm, and injure skin and other tissues. A typical case of cold injury occurred in Fairbanks, Alaska, in 1965. The official *maximum* temperature was −40°C (−40°F) or lower for

17 days. A woman, whose car had stalled, walked for 15 minutes with her legs covered only by silk stockings, and subsequently spent a month in the hospital with frostbitten legs.[3] This exemplifies the fact that in our modern age of heated homes, automobiles, and businesses, the average person is not aware that all of his or her physiological reserves for combating cold may suddenly be drawn on.

In the cold climates of earth, there are multiple environmental stressors that disturb homeostasis in the body's systems. The most obvious stressor is air temperature that is below skin and core body temperature. As figure 3.1 illustrates, the balance of heat production (e.g., from basal metabolic rate, exercise) versus heat loss (e.g., from evaporation, radiation, convection; see chapter 2) is disrupted; heat loss exceeds production and the central body temperature decreases, sometimes to the point of death by hypothermia (rectal temperature approximately 22°C, 72°F) or injury by frostbite (skin temperature –2 to 0°C, 28 to 32°F). The second common environmental stressor is air movement across the body. Wind greatly accelerates the loss of body heat, by (a) removing (via convection) still warm air layers trapped in insulative clothing, (b) increasing evaporative cooling when the insulative material is wet, or (c) increasing evaporative cooling directly from sweat-soaked skin.[4] Antarctic explorer Paul Siple coined the term "wind chill" to define the combined effect that low air temperature and air movement have on skin. This effect may account for as much as 80% of all heat lost from the body in cold air.[5] Figure 3.2 presents the widely recognized windchill chart, which describes the risk of freezing exposed flesh in cold environments.[6] You should note that flesh may freeze in one minute or less, if the windchill is severely low (in the "increased risk" or "great risk" categories). The risk of whole-body hypothermia also increases as the windchill becomes more severe.[7]

The third environmental stressor that disturbs the homeostasis of the body's systems and organs is dry air. Air with a very low relative humidity has been identified as a stressor at two sites: the hot, dry desert winds known as the Foehn or Sharav (see chapter 7), and cold, dry air that contains pollutants (see chapter 6). In this chapter, the effects of dryness plus cold air will be considered as they affect the smooth muscle surrounding lung and bronchial airways. Inhalation of cold, dry air stimulates exercise-induced bronchospasm (EIB), whereas warm, moist air (e.g., during swimming) is tolerated much better by asthmatics and other sensitive individuals.[8] This effect apparently occurs even when warm, dry air is inhaled by healthy adults at high ventilatory rates; in this instance, FEV_1 is reduced.[9]

Water immersion, of the entire body or the extremities, is the final environmental stressor that may be associated with cold

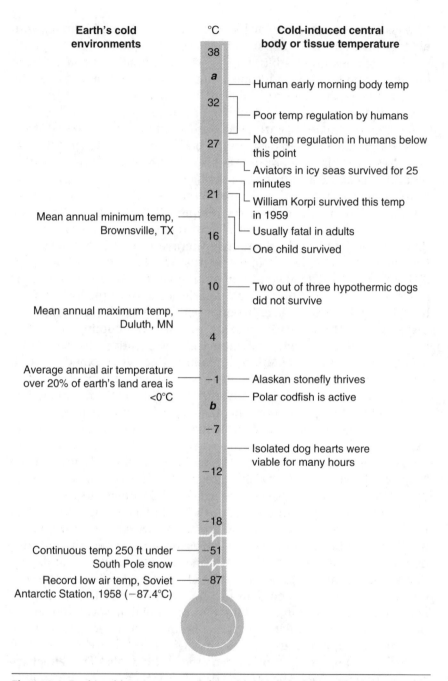

Earth's cold environments	°C	Cold-induced central body or tissue temperature

- 38
- *a*
  - Human early morning body temp
- 32
  - Poor temp regulation by humans
- 27
  - No temp regulation in humans below this point
  - Aviators in icy seas survived for 25 minutes
- 21
  - William Korpi survived this temp in 1959
- Mean annual minimum temp, Brownsville, TX
- 16
  - Usually fatal in adults
  - One child survived
- 10
  - Two out of three hypothermic dogs did not survive
- Mean annual maximum temp, Duluth, MN
- 4
- Average annual air temperature over 20% of earth's land area is <0°C
- −1
  - Alaskan stonefly thrives
  - Polar codfish is active
- *b*
- −7
  - Isolated dog hearts were viable for many hours
- −12
- −18
- Continuous temp 250 ft under South Pole snow — −51
- Record low air temp, Soviet Antarctic Station, 1958 (−87.4°C) — −87

Figure 3.1 Earth's cold environments (left) and cold-induced central body or tissue temperature in various animals (right): *a* hypothermia is defined as a rectal temperature below 36°C; *b* frostbite occurs when skin temp falls below –2 to 0°C. Adapted from Folk 1966.

Air temperature	Estimated wind speed in mph (kph)				
	0 (0)	10 (16)	20 (32)	30 (48)	
30°F (−1.1°C)	30 (−1.1)	16 (−8.9)	4 (−15.6)	−2 (−18.9)	Little risk
20°F (−6.7°C)	20 (−6.7)	4 (−15.6)	−10 (−23.3)	−18 (−27.8)	
10°F (−12.2°C)	10 (−12.2)	−9 (−22.8)	−25 (−31.7)	−33 (−36.1)	Increased risk
0°F (−17.8°C)	0 (−17.8)	−24 (−31.1)	−39 (−39.4)	−48 (−44.4)	
−10°F (−23.3°C)	10 (−23.3)	−33 (−36.1)	−53 (−47.2)	−63 (−52.8)	
−20°F (−28.9°C)	−20 (−28.9)	−46 (−43.3)	−67 (−55)	−79 (−61.7)	Great risk

Figure 3.2 Windchill chart: the risk of freezing exposed flesh.
Adapted from Milesko-Pytel 1983.

environments. In contrast to wet skin, hair, or clothing, which accelerates evaporative cooling, water immersion (which involves convection and conduction) may cause faster cooling of core body temperature than any other environment that laborers, soldiers, and athletes face. At rest, the fact that cooling via cold water is 20-25 times greater than in air at the same temperature[10] explains why (a) pilots typically die from **hypothermia** when they perform emergency landings in the ocean (see figure 3.1), and (b) ice-water immersion is the most effective whole-body cooling technique (see bars f and g in figure 2.11, page 55). Therefore, whenever working or exercising in a cold environment, you should avoid water immersion.

YOUR BODY'S RESPONSES TO COLD

Winter sports and recreational activities draw more people outdoors than ever before.[11] Alpine and cross-country skiing, climbing, backpacking, hiking, snowboarding, and snowmobiling all expose the body

The Body's Responses to Stress

Nervous System

Maintains homeostasis:

- hypothalamus receives afferent input
- sends efferent messages that maintain/alter body temperature and regulate cutaneous blood vessel responses

Muscular System

- shivering produces internal metabolic heat

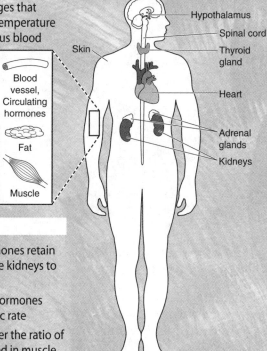

Endocrine System

- fluid-electrolyte hormones retain water and NaCl via the kidneys to offset dehydration
- thyroid and adrenal hormones increase the metabolic rate
- adrenal hormones alter the ratio of metabolic fuels utilized in muscle

Cardiovascular System

- cutaneous blood vessels constrict; this retains heat in the body and maintains central blood volume and pressure
- heart rate and cardiac output are regulated

Subcutaneous Fat

- adipose tissue retains heat by insulating the body

Urinary System

- kidneys retain water and NaCl to offset sweat losses

Figure 3.3 Organs and systems that respond to cold exposure and protect tissues from injury.

to cold air and windchill. Even during spring and summer months, hiking in cool mountain air during a rain shower can result in a rapid drop in central body temperature (see "Hypothermia," page 106). During these outdoor activities, various **adaptive responses** occur in numerous organs, as shown in figure 3.3, due to interactions between sensory receptors (those that collect information about the location, nature, and extent of homeostatic imbalances), the central nervous system (CNS), and organs (e.g., skeletal muscles, blood vessels, the thyroid gland, and the adrenal glands). Each organ plays a unique role in accomplishing the goals of this activated state: to maintain core body temperature at or near 37°C (98.6°F) and to mobilize fuels (e.g., blood glucose and fatty acids) that sustain increased metabolism and heat production (figure 3.4, page 78).

Before you examine figure 3.4 further, you may want to review two sections in previous chapters.[11, 12] These will refresh your memory about the physiological mechanisms involved, will enhance your understanding of the adaptive responses that occur, and will likely allow you to appreciate the integration that is accomplished by the CNS. The first section, "Principles Regarding Exposure to Stressful Environments," is on page 4 in chapter 1, and the second, "How Your Body Loses Heat," appears on page 17 in chapter 2.

Adaptive Responses

Focus for a moment on the two columns on the right side of figure 3.4, labeled "Organs involved" and "Responses." It soon should become obvious that a variety of tissues work in your body to offset heat loss in a cold environment. These tissues include nerve, brain, skeletal muscle, smooth muscle, blood vessel, and endocrine glands, plus at least five distinct hormones.[11] Your body's approach to cold stressors (cold air, windchill, cold-water immersion) is three-pronged: increase heat production, reduce heat loss, and mobilize metabolic fuels. Heat production increases when the rhythmic muscular contractions of shivering begin. **Shivering thermogenesis** is stimulated when heat loss causes your skin temperature to fall rapidly or your core body temperature to fall to the shivering threshold (labeled CT in figure 3.5, page 79),[13-15] which is the point to which body temperature must fall before shivering begins. Because these contractions do not produce any work, the energy is mostly converted to heat. Heat loss is reduced when your cutaneous blood vessels constrict, thereby lowering the volume of warm blood flowing through the skin. Metabolic fuels (e.g., free fatty acids and glucose) are released into your bloodstream when the hormones **epinephrine** and **norepinephrine**

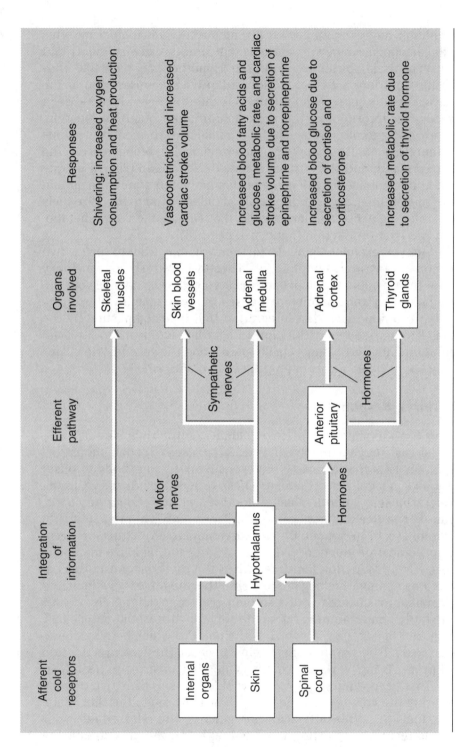

Figure 3.4 Adaptive responses by which your body regulates its core temperature in a cold environment.
Adapted from Noble 1986.

are secreted from the adrenal medulla, and when cortisol and cortico-sterone are secreted from the adrenal cortex. And, your overall meta-bolic rate (heat production) increases when the thyroid gland secretes thyroid hormone. During most exposures to cold air, the first response is peripheral vasoconstriction, followed by shivering and hormone release.[16] This skin vasoconstriction lowers skin temperature (T_{sk}), which in turn reduces heat loss via radiation.

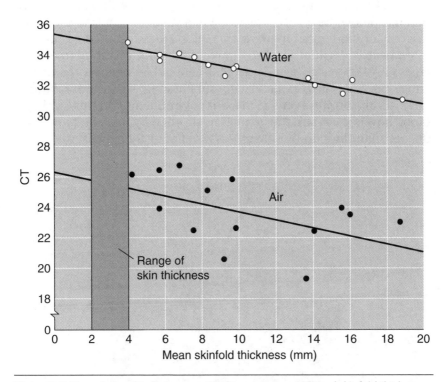

Figure 3.5 The relationship between critical temperature (CT) and skinfold thickness, a measure representing the amount of subcutaneous fat. CT is the water/air temperature at which shivering begins and metabolic rate increases.
Adapted from Smith and Hanna 1975.

The effectiveness of skin **vasoconstriction** in reducing heat loss (conserving body heat by shunting warm blood into deep veins) is influenced by the amount of adipose tissue (fat) stored under the skin. This means that heat loss is inversely related to the mass of subcutaneous fat; as fat mass increases, heat dissipation decreases. This is due to the fact that fat and skin insulate the body when skin blood vessels are tightly constricted.[13] In fact, fat is an excellent insu-lator in water and in air (figure 3.5) that allows T_{sk} to decrease with a concomitant drop in radiative (air) and convective (air, water) heat loss. This explain (a) why persons with more subcutaneous fat can tolerate a lower water or air temperature than lean persons before shivering begins;[15] and (b) why metabolic rate is inversely related to body fat content during cold exposure.[17] However, because skin blood vessels pass through, and are distally superficial to, the subcutaneous fat layer (see figure 3.6), these relationships exist only if vasoconstriction is maximal or nearly maximal, because the flow of warm blood is below the level of the fat. When vasoconstriction is mild (as in a cool or moderate environment), blood flow to skin is greater and more heat is dissipated to the surrounding air, resulting in a lower core body temperature and a greater demand for heat production via shivering.[18] This also explains why subcutaneous fat is not related to heat storage during exercise in a hot environment.[19] In this case, skin blood flow is great and the insulation afforded by body fat (which lies deeper than the skin blood vessels) is effectively bypassed.

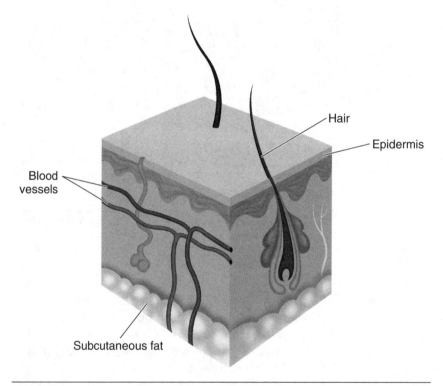

Figure 3.6 Anatomical relationship between blood vessels, subcutaneous fat, and skin.

As described in chapter 2 (see "Cardiovascular Responses to Heat and Humidity," page 22), cardiac output (liters per minute) has two components: stroke volume (liters per beat) and heart rate (beats/minute), both of which are increased in a hot environment. However, a resting exposure to cold air (5°C, 41°F), which increases the demand for blood flow and oxygen delivery to shivering muscle, increases cardiac output solely by an increased stroke volume while heart rate remains constant.[12, 16] Two theoretical mechanisms have been proposed.[16] It is possible that elevated circulating catecholamine levels increase the stroke volume. It also is possible that pressure receptors (e.g., baroreceptors) sense an increase in central blood volume following peripheral vasoconstriction, which depresses heart rate via a reflex. This adaptive response strikingly contrasts with the mechanism employed during exercise in cold air (see below). A resting, unclothed exposure to cold air also increases average blood pressure.[12]

All of the adaptive responses shown in figure 3.4 are under autonomic nervous system control. However, behavior is an important, complex aspect of temperature control in humans and has been named **behavioral thermoregulation**.[20] Simple examples of this phenomenon, which occurs when a person senses that he or she is cold, include adding items or more layers of clothing, drinking hot liquids, and moving to a warmer location (such as indoors). In nonhuman species, behavioral thermoregulation is mostly limited to instincts. The insulative value of clothing will be considered in detail in the final section of this chapter.

In terms of body fluid balance during cold-air exposure, a detrimental response arises indirectly from skin vasoconstriction (i.e., reduction of peripheral blood flow) and the resulting relative increase of central blood volume. Volume receptors in the left atrium of the heart inhibit the secretion of the hormone AVP (see chapter 2 section titled "Urine: An Index of Fluid Balance," page 33) from the pituitary gland in the brain.[13] In the kidneys, the amount of water reabsorbed into the blood is controlled in part by AVP. This means that skin vasoconstriction indirectly results in an increased urine volume during cold exposure. It is difficult to categorize this response as beneficial because it accelerates dehydration, thereby affecting physical performance and health negatively. The impact of dehydration on soldiers and laborers who work in cold environments will be described in the final pages of this chapter.

As with heat and air pollutants, there is evidence to suggest that the human physiological responses to cold-air exposures at rest alter with age. For example, it has been shown, in a comparison of elderly

© Betty Crowell

Exposure to cold triggers an array of adaptive responses designed to maintain core body temperature.

men 69-70 years versus younger men <45 years, that aging reduces skin blood flow and the degree of vasoconstriction in response to cold-air exposure.[21-23] Research also has shown that the cold-induced increase in resting metabolic rate (see figure 3.4) is smaller in older than in younger subjects.[22, 24] This causes older individuals to experience a lower rectal temperature while resting in a cold (5°C, 41°F) environment, because metabolic heat production is lower.[22]

Energy Expenditure and Metabolism

The total energy expenditure of your body is relatively constant from day to day and consists of three major components: resting metabolic rate (RMR); the energy cost of food absorption, metabolism, and storage (F); and the energy cost of exercise and other movements (EX). RMR accounts for the largest portion of daily energy expenditure (60-75%), and is affected by many intrinsic and extrinsic factors, including age, fat-free muscle mass, prolonged exercise, clothing, and ambient temperature. This final factor is relevant to the present chapter because it may or may not alter RMR, depending on the severity of the environment.[25] Figure 3.7 illustrates this fact; it was published by Eric Poehlman and colleagues at the University of Vermont.[26] This figure shows that there is a zone in which ambient temperature changes do not alter RMR appreciably (i.e., at the lowest point in the curve). When air temperature is considerably lower than core body temperature, RMR increases, via involuntary muscular activity (e.g., shivering) and hormonal effects, to replace the heat that was lost to the environment (see figure 3.3, page 76). Figure 3.7 also allows you to mentally compare the energy expended during shivering at rest to typical energy expenditures during exercise (i.e., 100 kcal/mi of running or walking).

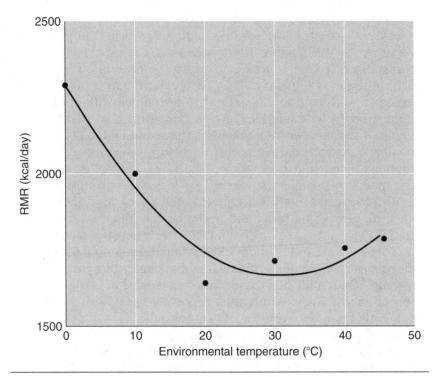

Figure 3.7 The influence of environmental temperature on resting metabolic rate (RMR).
Adapted from Poehlman et al. 1990.

The nutrients that are used by cells for fuel (metabolic substrates) during rest or light daily activities also change in cold environments. In mild conditions, lipids are the primary substrates, accounting for 59% of all energy expenditure. When resting for 2 h in cold air (5°C, 41°F), carbohydrates become the preferred fuel for shivering thermogenesis, derived from both blood glucose and intramuscular glycogen stores.[27] As you will see in the next section of this chapter, these responses are different during exercise in cold air. Interestingly, the onset of low blood sugar (hypoglycemia) has been shown to inhibit shivering thermogenesis in humans;[28] this further emphasizes the importance of carbohydrates as a substrate for shivering.

COLD AIR AND PHYSICAL PERFORMANCE

Numerous difficulties arise from exercise or work in cold air; all these are stressors involving either insulative clothing or body cooling.[29] The first involves the awkwardness and extra weight of protective clothing. As the weight of clothing increases, caloric expenditure increases at a rate that is greater than can be accounted for by the weight of the clothing alone. This extra caloric output has been attributed to both a "hobbling effect" (interference with joint movements caused by bulky clothing) and "frictional drag" caused by layers of material sliding over each other.[16, 30] The second difficulty involves shivering, or an increase in muscle tone before shivering begins. Both conditions increase the metabolic cost (e.g., $\dot{V}O_2$) of exercise at a given speed; this means that exercise is less efficient during shivering. Also, because shivering occurs in both agonists and antagonists, normal movement patterns and muscular coordination can be impaired.[29] Third, intense skin cooling effectively anesthetizes sensory receptors in the hand. This markedly impairs manipulative motor skills requiring finger dexterity, such as marksmanship, catching, and throwing.[29] One published experiment reported that manual dexterity declined when skin temperature fell below 13°C (55°F).[31] Although not directly related to exercise, two other difficulties arise from prolonged cold-air exposure:[32] (a) tissue injury to skin from frostbite (see below), and (b) the negative psychological and emotional effects (e.g., impaired cognition and clinical depression) that develop during prolonged labor and daily activities in a cold climate.

Controlled laboratory studies have evaluated differences between physiological responses to exercise in cold air and those in thermoneutral conditions. Table 3.1 summarizes the findings of these studies with regard to specific thermoregulatory, cardiopulmonary, and metabolic measurements.[33] However, it is important that you realize that this table presents a generalized statement about each variable, and that a cluster of physiological responses, such as those in table 3.1, cannot describe *all* exercise responses to *all* cold conditions

Table 3.1

A Comparison of Physiological Responses That Occur During Exercise in Thermoneutral (TN) and Cold (C) Environments[34, 35]

Variable	Thermoneutral[a]	Cold[b]
Thermoregulation		
Respiratory heat loss	Increases with increased $\dot{V}_E$	Greater than TN
Peripheral heat loss	Increases with workload or time	Greater than TN
Muscle temperature	Increases	Lower, similar to TN
Core temperature	Increase or no change	Decreases, similar to TN
Threshold temperature for sweating	No change	Less than TN
Cardiopulmonary changes		
Ventilation	Increases with increased workload	Higher than TN (hyperventilation)
Oxygen consumption	Increases with increased workload	Higher than TN for lower workloads; similar to TN for high workloads
Heart rate	Increases with increased workload	Lower than (or similar to) TN
Cardiac output	Increases with increased workload	Similar to TN
Peripheral blood flow	Increases with workload or time	Skin depends on exercise intensity; decrease (or no change) in muscle
$\dot{V}O_2$ max	No change	Less than TN

(continued)

Table 3.1
(continued)

Variable	Thermoneutral[a]	Cold[b]
Metabolic effects		
Free fatty acid use	Gradual increase	Less than TN
Glucose use	Higher at start; gradual decrease with time	Slightly higher than (or similar to) TN
Rate of muscle glycogen use	Increases with increased workload	Higher in cool muscle
Lactate production	Increases with increased workload	Higher than TN

[a] Conditions of light-moderate workloads, mild ambient temperature, and steady-state exercise.
[b] Relative to that of the thermoneutral condition.

Adapted from Doubt 1991.

(ranging from mild windchill to severe windchill; see figure 3.2, page 75). The interactions of exercise and cold stressors are complex, and are not necessarily additive or synergistic, as might be expected in hot (chapter 2) or polluted (chapter 6) environments. In fact, as exercise intensity increases, the differences in response to thermoneutral and cold environments may change. Therefore, if you initially acknowledge the complexity of exercise-cold responses in humans, it will be beneficial for you to consider the following generalized description of a common exercise scenario.[33, 16] This description parallels table 3.1.

Imagine that you have just left your home for a 3 mile jog on a cold, windy day in January. You did not dress warmly, so this exposure to exercise in cold air will provide an obvious challenge to the homeostatic balance of your body. After stretching for a few minutes, you begin to run. The cold air immediately decreases your skin temperature and your cutaneous vessels constrict strongly. Blood is diverted away from the skin to your central veins, and stroke volume increases due to increased venous return to the heart. Cardiac output and heart rate are thus maintained at levels below what they would be in a warmer environment.[16, 33] Vasoconstriction in skin (and perhaps in muscle), coupled with a lower muscle temperature, may increase the rate of anaerobic metabolism[34, 36] (e.g., glycolysis) and may reduce the clearance rate of metabolic by-products, such as lactate, from your muscles.[34] This effect is dependent on exercise intensity, as witnessed by changes in plasma lactate.[37] Verifying this increase of anaerobic metabolism, the rate of muscle glycogen utilization during your winter jog is greater than if you were running in the summer.[36] Interestingly, research shows that you might utilize more fat as a fuel if you exercise in a cold environment, such as 60 min of continuous cycling at an intensity of 66% of maximal heart rate in an air temperature of $-10°C$ ($14°F$),[38] or 60 min of continuous cycling (50 W) in an air temperature of $10°C$ ($50°F$).[39] However, it is fair to note than some studies disagree that this will occur,[29, 33] and that other studies suggest that *both* your carbohydrate and fat utilization will increase, with the former serving as the predominant fuel source, as occurs during rest.[27]

Increased shivering, which occurs concurrently with exercise, increases your $\dot{V}O_2$ above that experienced in a mild or warm environment.[16, 34, 40] Because shivering does not produce external work, the energy utilized in muscle is converted almost entirely to heat. This warms internal organs and raises the core body temperature slightly, but the amount of shivering-induced heat production is quite small

© Aneal Vohra/Unicorn Stock Photos

Inadequate clothing compromises your body's ability to maintain homeostasis.

in comparison to that produced by exercise. Shivering during exercise occurs only when your T_{re} falls below approximately 36°C (97°F).[11]

Your T_{re} will not be maintained in cold air if you jog at a slow pace. At higher exercise intensities, it will increase.[13] This will send conflicting stimuli to your brain, which will decrease shivering and sympathetic vasoconstriction, encouraging *heat loss* while cold receptors in the skin simultaneously signal the hypothalamus to *conserve* heat.[16] Although your overall heat loss will increase during intense running, the additional metabolic heat production will prevent a decline in rectal temperature.[33] In contrast, you also may begin to sweat as your core body temperature rises, even though your skin is relatively cool; wearing sweat-soaked clothing on a windy day may result in hypothermia (see section below titled "Counteracting Cold and Windchill," page 116). This explains why you should wear layers of quick-drying fabrics near your skin and have dry clothing on hand. This also illustrates why it is difficult to generalize about your body's responses to exercise in cold air: the physiological responses depend on the interactions of many factors.

In addition to maintaining rectal temperature(T_{re}), it also is important to maintain skin temperature (T_{sk}) when the windchill is severe (figure 3.2, page 75). Increased blood flow, due to skin vasodilation and increased cardiac output, can warm peripheral tissues in your fingers, toes, ears, and nose, protecting them from frostbite. This response is related to exercise intensity, according to studies that involved mild to moderate exercise in a 5°C (41°F) environment.[18] When light exercise was performed, T_{sk} cooled to 12°C (54°F);[41] subsequent moderate exercise caused a rapid rise in T_{sk} to between 20 and 34°C (68-93°F).[42] This effect of increased exercise intensity on local blood flow can benefit the peripheral areas of your body that are prone to frostbite.[18]

Heat loss through your lungs also modifies the thermal balance of your body's core, but only to a small degree (9% of total heat production).[43] The nasal sinus and pharynx region inside your skull warms air very rapidly and effectively. Large differences in the temperature of air inhaled during exercise, such as 24°C (75°F) versus –35°C (–31°F), cause virtually no change in rectal temperature.[44]

Effects of Physical Training and Aging on Responses to Cold Air

Physiological responses to prolonged cold air exposure (1-10°C, 33-50°F, for 2 h) are related to one's physical fitness level. Fit subjects have more efficient thermoregulatory abilities than unfit subjects, as shown by greater metabolic heat production and insulation of the body's core via constriction.[45] Apparently, physical training enhances the sensitivity of the adaptive responses shown in figure 3.4 (page 78). For example, T_{re} decreases to a lower level in trained distance runners (versus sedentary control subjects) before shivering begins.[46]

Physical training indirectly improves an athlete's ability to offset the stress of cold air or windchill, in two ways.[47] First, the increased muscle mass, which results from consistent training, provides greater metabolic heat production during shivering than in sedentary individuals.[18] Second, metabolic heat production (indirectly measured as $\dot{V}O_2$) can be sustained longer, because a high exercise intensity can be maintained by a highly trained endurance athlete who has a very high maximal aerobic power ($\dot{V}O_2$max).[47] However, this does not mean that all researchers have observed these responses. Horvath, for example, reviewed the findings of ten early reports and summarized them by recognizing that a few studies found no relationship between physical fitness and temperature regulation in the cold.[48] He suggested that this may have resulted from differences in the nature of the stressors presented to subjects, the type and magnitude of the training programs, or the inability of experiments to separate effects of training from the effects of repeated cold exposures prior to the final exercise test.[48]

The natural process of aging alters one's ability to regulate body temperature. As noted earlier in this chapter, these effects appear during rest as a blunting of the vasoconstrictor response to cold air exposure[21, 22] and as a reduction in shivering-related heat production.[22, 24] Further, during exercise, older age is associated with a faster decrease of T_{re} (e.g., in 30 min cycling in 5°C [41°F] air temperature) than that experienced by young adults. Among older individuals (60 and 64 years average age), it is noteworthy that a higher $\dot{V}O_2max$ does not afford protection against this decline in body temperature. This indicates that aging per se, and not physical fitness or $\dot{V}O_2max$, causes this decline in thermoregulatory ability.[22] One plausible explanation involves vasoconstriction, especially in the thigh region. During cycling exercise, when blood flow to the quadriceps muscles is elevated, it is possible that skin blood vessels cannot maintain constrictor tone effectively. This would result in increased skin blood flow and increased heat dissipation to the surrounding cold air. Skin temperature measurements and calculations of thermal conductance through the skin of older adults support this paradigm.[22]

Exercise Performance and Cold Air Exposure

Several sports are contested outdoors during winter months, including ice hockey, road or cross country footraces, snowboarding, alpine or cross-country skiing, and ice-skating. Other sports and recreational activities sometimes occur when the air temperature is cold; these include American football, soccer, baseball, hiking, and rock climbing. Performance in these sports and activities may be negatively affected if deep body temperature decreases.[13]

Because preexercise *warm-up* prepares muscles and the cardiovascular system, helps to prevent injury, and promotes metabolic efficiency,[49] it is important that you understand a study conducted in Japan in 1992. The purpose of that research was to observe physiological responses to different warm-up protocols when ambient temperature was 10°C (50°F).[50] The test subjects were trained skiers who performed five warm-up runs on different days: 15 min of exercise at 70% $\dot{V}O_2max$; 15 min at 50% $\dot{V}O_2max$; no warm-up; 30 min at 70% $\dot{V}O_2max$; and 30 min at 50% $\dot{V}O_2max$. Test results showed that the optimal preparation for exercise occurred after the latter test, involving moderate exercise intensity and longer duration. Warming up with a short duration, high-intensity protocol was ineffective. The following paragraphs describe other physiological studies that complement this finding.

Exercise and work performance, as well as acclimatization to cold (see below), depend on more than the temperature of the air or water surrounding the body. If several layers of insulative clothing are worn, for example, it is possible that core body temperature might even *rise* during exercise in the cold. Therefore, regardless of the severity of cold exposure, you should focus on the effects on physical performance of a fall in core temperature or muscle temperature.

Cardiorespiratory endurance will likely decrease if the whole body cools, and this is more likely to occur if exercise in cold air is prolonged.[13] Laboratory studies indicate that this decrease in endurance performance is due to at least three distinct adaptive responses, which may differ depending on the environmental conditions and exercise protocol utilized. First, maximal heart rate decreases with body cooling.[51] Because stroke volume does not increase, cardiac output falls (cardiac output = heart rate × stroke volume; see chapter 2, page 22), as does the ability to sustain strenuous exercise. Second, when blood temperature decreases below 37°C (98.6°F), less O_2 is delivered to skeletal muscle and other tissues because hemoglobin's molecular structure binds to O_2 molecules more tightly.[52] When both cardiac output and the dissociation of oxygen from red blood cells fall, $\dot{V}O_2$max decreases. Third, blood flow to skeletal muscle during exercise decreases when the body is cooled. Because this reduces O_2 delivery to muscles, more energy must be produced via anaerobic metabolism, increasing lactate levels in blood and muscle.[51] Of these three adaptive responses, the first has the greatest impact.

Interestingly, a recent investigation conducted in Scotland demonstrated that cardiovascular endurance performance in cold air declined, even when body temperature was elevated.[40] The top panel of figure 3.8 (page 92) depicts the influence of air temperature on the cycling time to volitional exhaustion of eight subjects. Clearly, the 11°C (52°F) air temperature provided the least stressful condition for endurance performance. The 21 and 31°C (70 and 87°F) environments decreased exercise time for the reasons described in chapter 2 (see "Heat and Physical Performance," page 24). Regarding the present topic, cooling the air from 11 to 4°C (52 to 39°F) caused a decrease of 12.1 min in cycling time to exhaustion. However, the bottom panel in figure 3.8 shows that the average rectal temperature was elevated to 39°C (102°F) at the end of the 4°C (39°F) trial and was higher in all other trials (i.e., 40°C [104°F] in the 31°C [87°F] environment). Thus, since body temperature did not decrease, the mechanisms of this decrease in performance were different from those described in the previous paragraph. The authors of this investigation proposed a few possible reasons for the decline in cycling endurance time, but could offer no definitive explanations based upon their data. For example, it is known that pulmonary ventilation ($\dot{V}_E$) and oxygen consumption ($\dot{V}O_2$) are affected differently by cooling the airways, the body's core, and the skin.[40] The effects of airway cooling during exercise depend on the intensity of cold-air stress; $\dot{V}O_2$ may increase, decrease, or remain constant.[37] If central body temperature decreases, $\dot{V}_E$ and $\dot{V}O_2$ fall, until shivering thermogenesis occurs; these factors subsequently increase. When

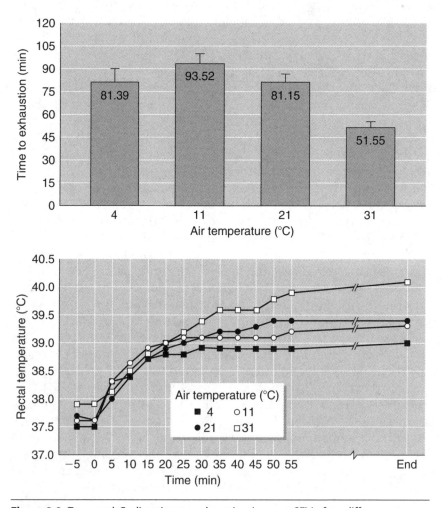

Figure 3.8 Top panel: Cycling time to exhaustion (mean ± SE) in four different environments. Bottom panel: Rectal temperatures recorded during each 55 min test. Adapted from Galloway and Maughan 1997.

the exercise intensity is great enough to maintain or elevate core temperature, but skin temperature decreases due to cold-air exposure, $\dot{V}_E$ and $\dot{V}O_2$ increase. Figure 3.9 illustrates this and complements figure 3.8, in that all data in these figures were collected during the same investigation.[40] Figure 3.9 demonstrates that $\dot{V}O_2$ increased the most, and that the skin temperature was lowest, during the 4°C (39°F) trials; yet, rectal temperature was elevated to a level that prevented shivering from occurring (figure 3.8). These increases in $\dot{V}_E$ and $\dot{V}O_2$ may have been caused by stimulation of temperature

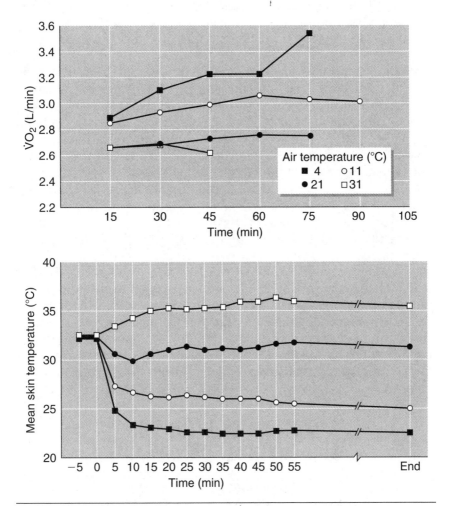

Figure 3.9 Top panel: Oxygen consumption ($\dot{V}O_2$) responses during exercise in four environments. Bottom panel: Mean skin temperature during exercise. Both figures correspond to those in figure 3.8.
Adapted from Galloway and Maughan 1997.

receptors in the skin, or increases in muscle tension and/or muscle metabolism.[53]

Muscular endurance, defined as the ability to sustain continuous contractions at submaximal intensity,[11] may be altered by cold-air exposure. As was the case with cardiovascular endurance, muscular endurance depends more on the decline in muscle temperature than changes in the surrounding air temperature. This effect also depends on the degree of cooling. When slight cooling of muscle is employed, fatigue develops at a slower rate.[54] However, when muscle temperature falls below 27°C (82°F), muscular endurance decreases.[13] This effect may be due to either reduced nerve conduction velocity or recruitment of fewer muscle fibers, especially those near the muscle surface.[55]

Maximal muscular strength and **peak muscular power** also decrease at lowered muscle temperatures.[29, 34] This negatively affects dynamic physical performance in virtually all sports and occupations. Specific decrements, subsequent to muscle cooling, occur in jumping and sprinting events.[56] The loss of performance in these events amounted to 4-5% per degree Celsius. There are several possible reasons why cooling may inhibit force production and power output.[13, 29] First, there may be an increase in the time it takes muscle fibers to reach maximal tension. This may involve a slower rate at which actin and myosin cross-bridges (i.e., the microfilaments in skeletal muscle) break and reattach.[57] Second, the viscosity of the fluid inside muscle fibers (sarcoplasm) may increase as the muscle is cooled, increasing the resistance to movement of the cross-bridges and actin.[13] Third, it is known that the rate of chemical reactions in muscle slows as the temperature drops, primarily because muscle enzyme activity and the production of high-energy phosphates (e.g., ATP) decrease.[29, 57] Due to these changes in maximal strength and power, the most susceptible types of exercise are those that are brief and dynamic, utilizing fast movement velocities and the elastic properties of active muscles.[58]

COLD-WATER IMMERSION AND PHYSICAL PERFORMANCE

The loss of exercise-induced body heat that occurs in cold air, and the adaptive responses made to counteract heat loss, are magnified and accelerated when the body is immersed in cool or cold water. This occurs because body heat is dissipated approximately two to four times as fast in cool water as in air, at the same temperature.[59] This is especially true for swimming, because the movement of water past the skin surface markedly increases heat transfer by convection.[18, 60] However, during exercise in cold water, skin heat loss (which involves conductance) can be 70 times greater than in air of equal temperature.[20] These facts support the use of water-immersion therapy to cool hyperthermic athletes and heatstroke patients, as discussed in chapter 2. You might review the rectal temperature cooling rates for air versus water exposure again, as shown in figure 2.10 on page 54.

Because heat loss from the surface of skin is so rapid during cool- or cold-water immersion, shivering is common and contributes a considerable amount of heat to the balance of core body temperature. In fact, the highest oxygen consumption during shivering represented about 50% $\dot{V}O_2$max in humans.[61] Similarly, **nonshivering thermogenesis** (i.e., heat production due to food metabolism or exercise) contributes to body heat balance, but is more difficult to measure than shivering. One measurement that physiologists have recorded to compare the body's adaptive responses in different watery environments is the **critical temperature** (CT), below which there is an increase in energy metabolism that exceeds the resting metabolic rate (RMR). You will recall that this concept was presented graphically in figure 3.5 (page 79), which illustrates that the CT is considerably higher than the critical air temperature required to increase metabolic heat production. In the database, the highest CT was approximately 35°C (95°F) for a lean subject who had little subcutaneous fat. For the majority of test subjects in that study,[15] however, the CT ranged from 30 to 34°C (86-93°F). This, obviously, is well above the temperature of ocean water in arctic climates (e.g., 4°C, 39°F).[62]

Even though the body begins to generate heat by shivering and nonshivering means, this is no guarantee that thermal homeostasis will be maintained. For example, it is known that a water temperature below 18°C (64°F) is too cold to sustain human life indefinitely.[63] It also is known that several factors besides water temperature determine survival in cold water. These include duration of immersion, body mass, activity level, position in the water (e.g., lying on the water surface or head-out immersion), clothing insulation, and subcutaneous fat.[62, 64] This latter factor is illustrated in figure 3.5, in that the statistical relationship between skinfold thickness (a measure of subcutaneous fat) and CT is very strong. Those humans with greater fat insulation begin generating heat, above the RMR, at a lower CT than their leaner counterparts. The increase in heat production is proportional to the decrease in water temperature, but contributes minimally to the maintenance of core temperature because cold water removes heat from the body so effectively.[18] However, in a cold water survival situation, extra subcutaneous fat can mean the difference between life and death.[62]

These facts have led scientists to create statistical models of survival time at different water temperatures. Typically, such models have been applied to pilots who have landed disabled aircraft in ocean waters. Figure 3.10 illustrates the probability of survival during water immersion. The solid lines delineate *safe*, *marginal*, and *lethal* zones, based on a combination of water temperature and immersion duration.[18] No deaths would be predicted in the safe zone. Under conditions described by the marginal zone, no deaths should occur due to hypothermia, but unconsciousness would result in 50% fatalities due to drowning. No human exposed to water immersion in the lethal zone would be expected to survive.[65] Although these predicted zones

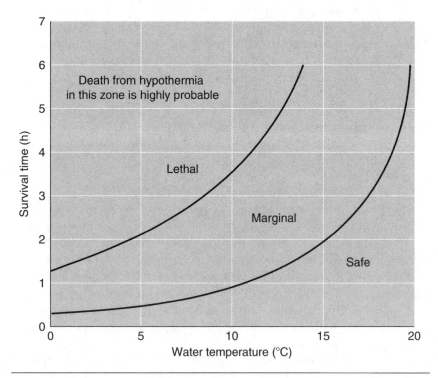

Figure 3.10 Estimated survival time during immersion of lightly clothed, nonexercising humans in cold water. The boundaries (solid lines) would be shifted upward by wearing survival gear or a wet suit.
Adapted from Toner and McArdle 1988.

of survival are conservative, and their boundaries perhaps could be shifted upward slightly, they represent a clear statement that water immersion is an obvious threat to life and health.[18, 66]

This fact is very relevant to commercial and recreational divers, channel swimmers, and triathletes, who spend long periods in cool or cold water. These individuals can experience considerable body heat loss, even in cool, comfortable water temperatures. However, if exercise is performed at a high intensity, it is theoretically possible that normal body temperature (normothermia) could be maintained as long as exercise continued. For example, studies indicate that leg exercise during water immersion (18 and 30°C, 64 and 86°F) helps subjects maintain core body temperature 0.4-0.9°C (0.8-1.6°F) higher than while resting in water.[67] In contrast, other studies have shown that rest is more effective than exercise during immersion. Two of these, published by Crittenden and colleagues[68] and Keatinge,[69] used water temperatures ranging from 5 to 20°C (41 to 68°F); convective heat losses due to the movement of exercise (i.e., from skin to water) were large. Indeed, survival guidelines have stated that individuals who are accidentally immersed in cold water should maintain a still posture, to conserve heat and maintain core body temperature.[68] Currently, this practical problem remains unresolved and warrants further research. It is likely that some environmental conditions would require exercise to avoid dangerous hypothermia, while other scenarios would require stillness.

Even the limbs that are involved in exercise can alter the rate of heat loss and influence rectal temperature. Figure 3.11, for example, illustrates the interactive effects that the type of exercise and water temperature had on ten test subjects who wore only a nylon swimsuit and were immersed to the first thoracic vertebrae. This graph demonstrates that leg exercise was more effective than either arm or combined arm-leg exercise in preventing the drop of rectal temperature, at three different water temperatures (20, 26, and 33°C; 68, 79, and 91°F). Physiologically, figure 3.11 is interpreted to mean that arm exercise results in greater convective and conductive heat loss than leg exercise in cold water. This may result from either less fat insulation on the arms, greater blood flow (per unit limb volume) in arms than in legs, or a smaller cross-sectional distance from core to surface in arms than in legs (less tissue to resist heat transfer).[64] When survival is an issue, as it is when pilots are forced to land in cold ocean water, this figure suggests that leg movements should be emphasized and arm movements minimized.

Metabolic fuels also have important implications in survival situations. Muscle glycogen, the carbohydrate storage molecule consisting

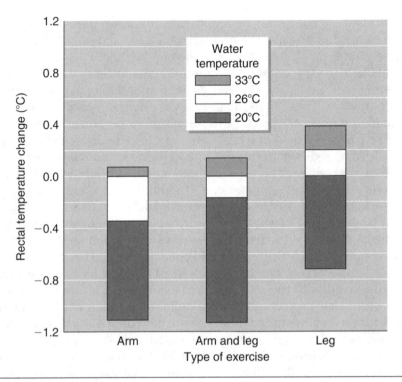

Figure 3.11 Change in rectal temperature after 45 min of water immersion. Exercise was conducted with different limbs and in three water temperatures.
Adapted from Pandolf et al. 1987.

of a chain of glucose units, is an important source of energy for human shivering thermogenesis[70] and, therefore, maintenance of body temperature. It is logical, then, that a low muscle glycogen concentration is associated with accelerated body cooling during cold-water immersion.[71] And, during prolonged shivering of many hours, it is possible that muscle glycogen depletion could occur, reducing the rate of heat production.[71] In muscle with normal glycogen levels, the duration of immersion required to affect shivering via glycogen depletion is longer than 3 h.[72]

In terms of athletic contests, it seems prudent[47] to limit endurance events so that the duration of water immersions do not exceed the predicted *safe zone* for a given water temperature (see figure 3.10). Further, athletes and laborers should be observed closely at the beginning of a contest or work shift, when water temperatures are uncomfortable or painful.[47] Plunging into very cold water can produce ventricular arrhythmias as well as gasping and hyperventilation; these latter two responses could result in water aspiration and drowning.[4]

Mental performance also may suffer as a result of immersion hypothermia. For example, a lack of mental acuity and task focus obviously will affect both work and athletic performance negatively. The medical literature indicates that the signs and symptoms of mild hypothermia (i.e., core body temperature of 32-35°C, 90-94°F) include confusion, apathy, withdrawal, and slurred speech.[73, 74] Mild immersion hypothermia adversely affects complex cognitive tasks, whereas simple ones remain unaffected. Specifically, central nervous system cooling is suspected to be the critical factor,[75] although the discomfort of cold-water exposure may cause psychological or emotional distress as well. Complete details regarding the clinical issues surrounding hypothermia appear later in this chapter.

During prolonged underwater labor, recreational diving, and endurance swimming events, it is customary to wear a wet suit for thermal insulation. However, at least two studies have shown that wearing a wet suit in moderately cold water does not prevent a decline of body temperature. For example, 1.5 h of water immersion (at 14°C, 57°F), performed by a wet-suited breath-hold diver, caused rectal temperatures to fall from 37.7 to 36.9°C (99.9 to 98.4°F).[76] Similarly, fin swimmers and underwater scooter divers, who propelled themselves through 17-18°C (63-65°F) seawater for 3.7 h, experienced a continuous decline of rectal temperature, despite wearing a full wet suit.[77] This decrease ranged from 0.3 to 1.3°C (0.5-2.3°F), with the lowest T_{re} measurement being 35.7°C (96.3°F), suggesting that clinical hypothermia (i.e., when symptoms are evident, internal body temperature below 34°C, 94°F) may be routine during exposures lasting more than 4 h in moderately cold water. This may be due, in part, to water leaking into dry suits that reduces insulation by as much as 50%.[78]

ACCLIMATIZATION TO COLD AIR AND COLD WATER

Since the late 1950s, numerous research studies have examined the responses of various ethnic groups that either inhabit cold climates or work in cold water.[79] The goal of these studies has been to identify how natives of Peru, Australia, Korea, Tierra del Fuego, and the Arctic Circle maintain body temperature during daily cold exposures. Some of these natives live in nomadic bands or live in subsistence economies with limited access to modern technology. Thus, depending on their nutrition, shelters, and clothing, they encounter different levels of whole-body and extremity cold stress.[79] For example, Inuits experience moderate whole-body cold stress because their clothing is so effectively insulated, with severe extremity chilling.

Australian Aborigines, in contrast, experience a prolonged moderate whole-body cold stress by sleeping in cool desert night air (e.g., 3-5°C, 37-41°F). These varied environmental conditions result in different adaptive responses. Inuits exhibit a high metabolic heat production and high levels of blood flow to the extremities, resulting in warm hands and feet.[80] The Aborigines oppose cold stress by constricting skin blood vessels (i.e., increasing peripheral insulation) and by tolerating mild hypothermia (e.g., 35-36°C, 95-96°F) without shivering-induced or nonshivering thermogenesis.[81]

Young's Model of Acclimatization

More recent investigations have identified five strategies the human body employs to adapt to cold. These strategies differ considerably. Table 3.2 differentiates the various responses that occur in each strategy. It should be obvious to you that the responses to cold exposure are not as consistent as those that occur during heat acclimatization (see table 2.2 on page 28). And, it is reasonable to wonder why so many different physiological strategies exist. One answer, for both cold-air and cold-water immersion, is that each adaptive pattern depends on the intensity of cold, the daily duration of the exposure, and the length of the acclimatization period. Another answer involves different scientific interpretations of data by investigators. As an example of this, physiologist Andrew Young has designed a theoretical scheme to explain the development of different patterns of cold adaptation that recognizes only three adaptive states: cold habituation, metabolic acclimatization, and insulative acclimatization.[82] His paradigm appears in figure 3.12, page 102. **Habituation** is a desensitization or dampening of the normal response to a stressor. A classic example of **cold habituation** was observed in fishermen who immersed their bare hands in cold water for many hours each day. Their skin vasoconstriction response was blunted locally by the central nervous system to allow increased cutaneous blood flow and a warmer skin temperature during their daily labor.[84] You may remember experiencing cold habituation if you have immersed your hand or body in very cold water. After a few minutes, you probably noticed that the pain and discomfort subsided. This response involved habituation of feeling from your pain-sensitive neurons in or near the skin (afferent information going to the brain). **Metabolic acclimatization** refers to change(s) in metabolism, either in the type of nutrient used for heat production or the metabolic pathway. **Insulative acclimatization** refers to change(s) in either skin blood flow or subcutaneous fat.

Table 3.2
Physiological Responses Observed Subsequent to the Five Types of Human Cold Adaptation. The Titles of Each Column Arise from Various Studies and Authors[82,83]

Measurments		Type of cold adaptation			
	Metabolic	Insulative	Hypothermic	Insulative-hypothermic	Metabolic-insulative
Rectal temperature	Normal (no change)	Normal (no change)	Decreases	Decreases	Normal (no change)
Skin temperature	Increases (vasodilation)	Decreases (vasoconstriction)	Normal (remains warm)	Decreases (vasoconstriction)	Decreases (vasoconstriction)
Metabolic heat production	Increases	Normal (no change)	Decreases	Increases	Increases

101

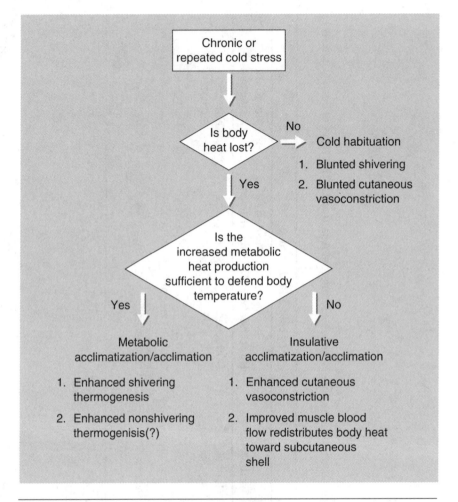

Figure 3.12 Flowchart illustrating a theoretical scheme to determine which type of human cold acclimatization/acclimation occurs.

From *Handbook of Physiology: Section 4: Environmental Physiology Two Volume Set,* edited by Melvin J. Fregly and Clark Blatteis. Copyright © 1996 by American Physiological Society. Used by permission of Oxford University Press.

One of the benefits of Young's flowchart (figure 3.12) is its simplicity when compared to the previous summary of many studies (table 3.2). Only two questions are posed: "Is body heat lost?" and "Is the increased metabolic heat production sufficient to defend body temperature?" A yes or no answer to these questions allows you to identify the type of acclimatization and the responses that occur. Another benefit of this flowchart is that it provides a summary of all of the major adaptive responses that the body can use: habituation of shivering and cutaneous vasoconstriction, enhanced shivering and nonshivering thermogenesis, and enhanced cutaneous or muscle blood flow. Unfortunately, three exceptions to Young's theoretical scheme exist. First, as noted in table 3.2, combinations of metabolic and insulative acclimation have been reported.[83] This apparently cannot be explained by figure 3.12. Second, a French research study has demonstrated that different people may adapt to cold-water immersion (10-15°C [50-59°F], 1-3 h/day, 4 days/week for 8 weeks) in different ways. Out of nine test subjects, one showed only metabolic acclimatization, three showed only insulative acclimation, and five exhibited a combination of these responses.[83] This suggests that either genetic or physical characteristics may alter the body's pattern of responses to cold exposure. In fact, Bittel interpreted his data with physical characteristics in mind, and found that physical fitness and percent body fat both influenced the type of adaptation that occurred. However, this was true only in those subjects who exhibited extreme responses. Third, it is theoretically possible that the different patterns of adaptation to cold exposure (e.g., metabolic, hypothermic, habituative, insulative) are actually progressive stages in the development of complete cold acclimatization.[82] This concept arose from observations of scuba divers in cold water, who initially responded by shivering but eventually lost this response and developed insulative adaptation.[85]

The Autonomic Nervous System and Cold Acclimatization

Another paradigm that explains cold acclimatization involves the autonomic nervous system. You will recall from chapter 1 that the two branches of the autonomic nervous system are known as the **sympathetic** and **parasympathetic** systems (see "Your Body's Responses to Stress," page 7). A study conducted by Mathew and colleagues in India found that humans initially responded to cold-air exposure (10°C [50°F], 4 h/day) with strong sympathetic responses.[86] These included shivering, cold-induced skin vasoconstriction, and elevation of blood pressure and heart rate during a cold-water hand-immersion test (HIT). After 21 days of this program, however, the parasympathetic nervous system predominated over the sympathetic, in that skin blood flow increased while shivering, pain, discomfort, blood pressure, and heart rate decreased during HIT. Thus, these investigators concluded that the shift toward parasympathetic predominance can be considered a characteristic feature of water-immersion cold acclimatization.[86]

The study of cold acclimatization is dynamically shifting each year. Until a paradigm is developed to explain most observations (like figure 3.12), authorities will continue to disagree. This situation has generated frustration in some researchers, who expected years ago that science would be able to explain cold acclimatization more clearly. One expert, for example, wrote that he wondered if, "it will be possible one day to produce the desired type of adaptation by making the correct choice of stress parameters. . . ."[87] This comment reflects the many uncertainties that remain in the scientific literature today.[83]

Endurance Training and Cold Acclimatization

Exercise performance is adversely affected by local effects of cold on the skeletal muscles, slowing of reflexes, and metabolic changes that alter the fuel supply to muscle tissue and the brain.[88] For example, the maximal contractile force of skeletal muscle is decreased by cooling.[89] The body may compensate for this by recruiting fast-twitch glycolytic muscle fibers, thereby using stored glycogen at a faster rate and hastening fatigue. The *hobbling effect* and *frictional drag*[16, 30] (see "Cold Air and Physical Performance," page 84) also are likely to induce glycogen depletion and early fatigue. Although hormonal changes (i.e., increased catecholamines, decreased insulin) favor lipids as an energy source, blood glucose levels may fall if a high exercise intensity is attempted after glycogen stores have been depleted.[70] Because the brain metabolizes only glucose, the judgment and skilled performance of athletes may deteriorate.[90]

© Jim West

The performance of athletes, such as this cross-country skier, can be greatly inhibited by exposure to cold.

As noted in table 3.2, insulative acclimatization is a normal response to repeated cold exposures. This insulative response results in a smaller loss of glycogen with a given cold exposure, because body temperature is higher and less shivering occurs, which has a correspondingly favorable effect on endurance performance.[90] Interestingly, the adaptive responses of cold acclimatization are similar to those of endurance training. There also have been suggestions of positive interactions between endurance training and cold acclimation.[83] Thus, regardless of the type of event that is contested, it is useful to include endurance conditioning in an athlete's program.[88] This type of training facilitates glycogen sparing by increasing the activity of fat-metabolizing enzymes and allows a competitor to exercise at a smaller percentage of $\dot{V}O_2max$.[90]

As noted on page 77, the **shivering threshold** is the point to which body temperature must fall before shivering begins. After repeated cold-air exposures (4-7 times over 2 weeks, 1 h duration, –5 to 5°C [23-41°F]), the shivering threshold is shifted to a lower body temperature. This allows the individual to tolerate cooler environments without

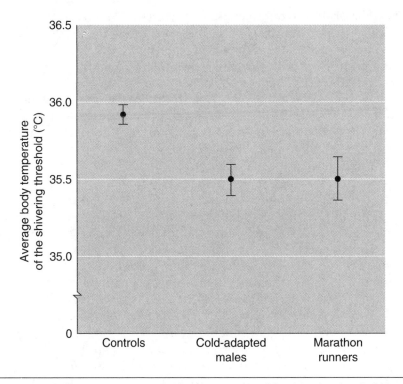

Figure 3.13 Differences in average body temperatures of the shivering threshold in control subjects, cold-adapted males, and marathon runners.[91]

wasted metabolic activity in skeletal muscles. As another example of a similarity between cold acclimatization and endurance training, the shivering threshold in marathon runners, who were not exposed to cold, also was shifted downward. Figure 3.13 (page 105) presents a visual comparison of the shivering thresholds of cold-adapted test subjects, marathon runners, and control subjects.[91] You will note that the thresholds of the two former groups are approximately 0.5°C (0.9°F) lower than the unacclimatized, untrained controls.

MEDICAL CONSIDERATIONS: COLD INJURIES

The four most common cold injuries are hypothermia, frostbite, immersion foot, and chilblains. The first two of these are medical emergencies that may eventually result in death. Only frostbite involves frozen tissue.

Hypothermia

The temperature cited for mild clinical hypothermia differs among experts, ranging from 32 to 35°C (95-96°F).[4, 74, 92, 93] The most commonly cited body temperatures, however, are 34 and 35°C (94 and 95°F).[4, 74] Some authorities also provide categories of hypothermia, such as those in table 3.3, to guide physicians, nurses, and EMTs with treatment. Perhaps the best explanations for these categorical differences are the (a) variability in symptoms and responses exhibited by patients, (b) patient characteristics (such as age, fitness, body fat content), and (c) differences in environmental conditions.

Hypothermia may occur during any season. Unexpected immersion in cold water, such as that experienced by capsized canoeists or whitewater rafters, is a common cause of hypothermia. Mountain bikers who encounter unexpected storms without proper clothing, unprepared glider pilots at high altitude, and open-water swimmers are also prime candidates.[92] Hypothermia also occurs among marathon runners who compete in 4-10°C (40-50°F) conditions. By the time they reach the middle of the race, they are hyperventilating and losing so much heat via radiation, convection, and evaporation that the body begins to cool very rapidly.[96] In the final miles—when exercise intensity and metabolic heat production wane—shivering, incoordination, and muscle spasms are often seen. In fact, the author of this book experienced this exact sequence of events in April 1986 during the Boston Marathon. The race began in sunny and warm conditions,

Table 3.3
Categories of Unintentional Hypothermia, as Published by Various Authorities

Clinical category	Deep body temperature (°C)			
	A	B	C	D
Mild	>32	>32	32-35	>32
Moderate	—	29-32	28-31	27-32
Deep/severe	<32	<29	25-27	<27
Profound	—	—	<25	—

Symbol: —, category not utilized.

A Departments of the Army, Navy, Air Force. 1985. *Cold injury.* St. Louis: U.S. Army Adjutant General Publications Center, Technical Bulletin 81, 1-14.
B Blue, B. 1994. Safe exercise in the cold and cold injuries. In *Sports medicine secrets,* edited by M.B. Mellion, 86-90. Philadelphia: Hanley and Belfus Co.
C Bangs, C., & Hamlet, M. 1980. Out in the cold—management of hypothermia, immersion, and frostbite. *Topics in Emergency Medicine 2*: 19-37.
D Danzl, D.F.; Pozos, R.S.; & Hamlet, M.P. 1995. Accidental hypothermia. In *Wilderness medicine. Management of wilderness and environmental emergencies.* 3rd edition, edited by P.S. Auerbach, 51-103. St. Louis: Mosby Year Book, Inc.

prompting competitors to wear only a T-shirt, shorts, and socks, but ended in cloudy, windy, and cool conditions. It was the perfect scenario for the development of hypothermia.

The warning signals of hypothermia are straightforward, and you should watch for them in yourself and in others. These include uncontrollable shivering; a grayish skin color; vague, slow, slurred speech; memory lapses; combative behavior; irritability; immobile fingers; staggering or stumbling; drowsiness; exhaustion; and inability to stand and move after a rest.[96, 97]

Several authorities have identified the factors that contribute to the lowering of body temperature in cool or cold air.[18] Two of the most important of these appear to be unanticipated wetting of clothing and extended length of exposure.[98] However, maintenance of a high metabolic rate during cold exposure combats the onset of debilitating hypothermia because the deep organs are heated. As exertion continues and fatigue grows, the intensity of exercise decreases; this lower rate of heat production contributes to the development of hypothermia.[99] Also, progressive hypothermia and the associated changes in coordination and consciousness probably contribute to fatigue,[18] because coordinated or skilled movements are required to maintain a high metabolic rate. In urban areas, most cases occur among alcoholics and drug abusers.[96]

The British thermal physiologist K.C. Parsons has equated the clinical signs of hypothermia with central body temperatures.[100] His work is summarized in table 3.4, which should be compared to table 3.3. As you read down this progressive list, you should realize that the

Table 3.4
The Clinical Signs and Symptoms of Hypothermia

Body temperature		
(°C)	(°F)	Clinical signs
37.0	98.6	"Normal" body temperature
36.0	96.8	Metabolic rate increases in an attempt to compensate for heat loss
35.0	95.0	Maximum shivering
34.0	93.2	Victim conscious and responsive, with normal blood pressure
33.0	91.4	Severe hypothermia below this temperature
32.0	89.6	Consciousness clouded; blood pressure becomes difficult to obtain
31.0	87.8	Pupils dilated but react to light; shivering ceases
30.0	86.0	Progressive loss of consciousness; muscular rigidity increases; pulse and blood pressure difficult to obtain; respiratory rate decreases
28.0	82.4	Ventricular fibrillation possible with myocardial irritability
27.0	80.6	Voluntary motion ceases; pupils nonreactive to light; deep tendon and superficial reflexes absent
26.0	78.8	Victim seldom conscious
25.0	77.0	Ventricular fibrillation may occur spontaneously
24.0	75.2	Pulmonary edema
22.0	71.6	Maximum risk of ventricular fibrillation
20.0	68.0	Cardiac standstill
17.0	62.6	No measurable brain waves (EEG)

Adapted from Parsons 1993.

point of maximal shivering (T_{re} of about 35°C, 95°F) represents the most effective defense that the body can mount. Then when T_{re} falls to about 31°C (87.8°F), shivering ceases, presenting the most significant clinical event. This is crucial because it means that the patient will not rewarm spontaneously and requires external heat for recovery.[92] Although large variations exist between individuals, almost anyone will become unconscious when T_{re} drops to 30-31°C (86-88°F). Below 30°C (86°F), a major risk arises due to a heart electrical conduction disturbance (e.g., ventricular fibrillation), probably caused by the direct effects of low temperature on the cardiac pacemaker.[92]

In contrast to the treatment for heatstroke, the goal of the initial treatment for hypothermia is to increase central body temperature to a normal level. The methods of rewarming involve either external (surface) or internal (core) procedures.[93] External heating results in the body's outer shell (skin and superficial tissues) being warmed before the inner organs, and may be accomplished actively (hot water bottles, immersion in 40°C [104°F] water, heating blankets) or passively in a warm room (with or without blankets). Internal rewarming refers to heating the central organs by using sophisticated surgical techniques such as gastric, thoracic, or peritoneal lavage (i.e., irrigation with warm saline solution), hemodialysis, and cardiopulmonary bypass. The latter technique utilizes a mechanical pump, a bubble oxygenator, and a heat exchanger connected to the femoral artery and the femoral vein in a circuit.[95] Because these methods are very sophisticated, nonmedical Samaritans are limited to external rewarming of hypothermic patients.

If you ever find yourself in a situation that requires you to warm a victim of unintentional hypothermia, you can utilize the following methods:[94, 95]

- Allow the patient to inhale warm air.
- Keep the patient dry and out of the wind.
- Wrap the patient in a sleeping bag or layer of blankets that is covered with an additional layer of insulating material.
- Insulate hot water bottles with an article of clothing and place them on the patient's underarm and groin areas.
- Huddle with the patient in close body-to-body contact.
- Provide radiant heat.
- Provide warm fluids if the patient is only mildly impaired, can swallow, and is alert.

However, care must be taken not to burn the patient with hot objects or hot air.[95] You also should recognize that external heat causes skin blood vessels to dilate, which may decrease blood pressure when dehydration exists. This is significant because hypothermia patients are often hypotensive (have low blood pressure).

The Alaska Department of Health and Social Services has developed a set of guidelines for recognition and treatment of unintentional hypothermia. This booklet is designed for the general public and medical technicians. It describes actions that a first responder should take when a victim of hypothermia is found in the field.[101] You might find it to be a useful addition to your library.

Frostbite and Frostnip

The simplest classification of **frostbite** involves three clinical categories: frostnip, superficial frostbite, and severe frostbite. All of these involve freezing of some part of the body. In the case of **frostnip**, only the outer layer of skin is frozen. This cold injury may occur from contact with cold metal, a supercooled liquid, or from exposure to severe windchill.[102] Frostnip may be similar to a first-degree burn. After the frozen tissue is rewarmed, it may become red and peel over the course of several days. As it heals, the appearance is similar to a sunburn. Frostnipped skin should be warmed immediately by placing the injured skin on a warm body area (i.e., cheek, armpit) or into warm water maintained at 38-42°C (100-108°F).[102]

Superficial frostbite injures the outer layer of skin plus some underlying tissues. Externally, it has a white, gray, or mottled appearance. It feels hard on the surface, but soft and spongy below. Blisters usually appear within 24 h. Superficial frostbite should not be treated by simple application of heat. Rapid rewarming should be employed via immersion in warm water (38-42°C, 100-108°F), because this technique is the most efficient and thorough. Water of this temperature feels lukewarm to the unfrozen, normal elbow.[97] Proper rewarming is critical to healing.[102]

Severe frostbite involves crystallization of fluids in the skin. These crystals pierce skin cell membranes and result in tissue death. Tissues that are subject to a slow rate of cooling (i.e., hours) develop ice crystals in the interstitial space between cells. Rapid cooling (i.e., within minutes) produces intracellular ice crystals, which are more lethal to cells.[103] Interestingly, damage to microscopic blood vessels in the skin (arterioles, venules, capillaries), subsequent to the initial freezing and thawing, may be the primary cause of tissue death. This damage involves the formation of blood clots (thromboses), fibrin deposits (a part of the blood clotting mechanism) in the walls of arterioles, poor delivery of oxygen to cells (hypoxia), and degeneration of the inner walls of blood vessels. These effects are shown in figure 3.14, and result from one or more of the following: constriction of vessels, thickening of blood, and a complete cessation of blood flow.[104] Once the injury to microscopic blood vessels occurs, inflammation sets in. Over time, tissue hypoxia increases, in a manner that resembles a burn wound. One research team, in fact, reported that the chemical mediators that are released after frostbite (prostaglandins E_2 and F_2a, thromboxane

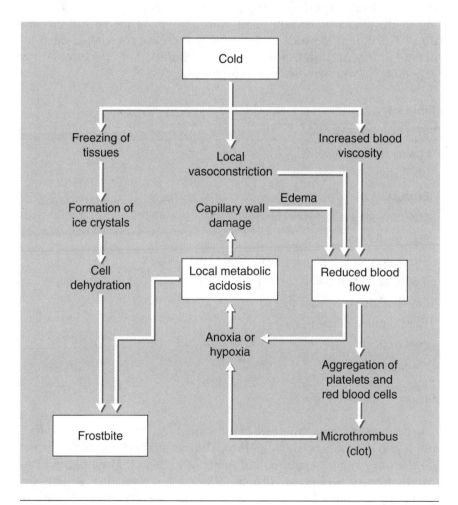

Figure 3.14 Mechanisms of frostbite.
Adapted from Foray 1992.

B$_2$) match those that appear after a third-degree skin burn.[105] They postulated that the massive swelling (edema) following cold injury was due either to the leakage of proteins through capillary walls (caused by the inflammatory mediators described above), white blood cells sludging inside capillaries, or increased hydrostatic pressure.[103]

The most common causes of frostbite and frostnip are presented in table 3.5. Three other predisposing factors[103] are not as widely recognized. First, cigarette smoking can cause vasoconstriction, decreased skin blood flow, and tissue death via frostbite, under severe environmental conditions.[106] Second, mental and physical fatigue render a person unlikely to perform simple preventive measures such as changing damp socks, massaging feet, and exercising to restore circulation.[4] Third, long periods of immobility increase the risk and extent of cold injury. Motion produces body heat and improves circulation, especially in wet, wind-exposed limbs.[4]

Table 3.5
Causes of Frostbite and Frostnip

Low air temperature

Wind chill

Exposed skin

Moisture on skin and in clothing

Poor insulation

Direct skin contact with a supercooled metal or liquid

Interference with circulation of warm blood

Cramped position or posture

Tight clothing or boots

Dehydration

Localized pressure

Adapted from Schimelpfenig and Lindsey 1991.

In civilian laborers, frostbite occurs most often on the toes and feet, but less commonly on the fingers and hands. It also has been reported on the ears, nose, scrotum, and penis in joggers.[94] Most patients recall that frostbite begins with cold hands and feet, after which pain gradually changes to numbness; the limb usually feels clublike or absent. While frozen, the extremity may appear pale and feel as hard as frozen meat. During rewarming, the loss of sensation is typically replaced with extreme pain. Within the first 24 h, swelling and blistering appear. However, it is initially difficult to predict the extent of frostbite damage because skin changes do not always reflect the damage to underlying blood vessels.[94] Only time allows an exact assessment of tissue death (necrosis). Table 3.6 presents the categories of frostbite injury based on observations made after freezing and rewarming. However, you should realize that there are obvious differences in responses among patients. In patients who experience no tissue loss, frostbite symptoms usually subside within one month. Those with tissue loss may experience localized disability for six months or longer. All frostbite patients experience some degree of sensory loss for at least four years after injury.[103]

Table 3.6
Frostbite Injury Categories and Their Associated Clinical Features[96, 106]

Injury category	Clinical features
First degree	Numbness, redness, edema; no tissue loss occurs; a firm white or yellow plaque occurs at the site of the injury.
Second degree	Blisters form and contain a clear or milky fluid; blisters are surrounded by redness and edema; solid black masses may form within 10 to 24 days.
Third degree	Deep blisters contain purple bloody fluid; this means that the entire skin thickness and blood vessels have been damaged; solid black masses form; healing occurs in an average of 68 days.
Fourth degree	Damage completely spans all layers of skin; this often causes mummification and gangrene with muscle and bone loss; a superficial line of healthy versus injured tissue forms in an average of 36 days and extends down to the bone 60-80 days after rewarming; part or all of the affected limb must be amputated or allowed to separate spontaneously.

Because the thawing (rewarming and refreezing) process appears to be a critical stage in the development of cell death, most experts recommend that you *not* apply first aid to victims of frostbite if there is a chance that the tissue will refreeze.[107] The safest action is to have the victim admitted to an emergency treatment facility. On admission, the frostbitten limbs will be rapidly rewarmed and a series of medical procedures will be instituted, including removal of blistered skin patches and administration of medications (for pain, inflammation, infection).[103] Anecdotal evidence and field experiences suggest that less tissue loss will occur during emergency transport if frostbitten limbs are not thawed, but left frozen until the patient can be treated in an emergency treatment facility.[102, 107-109] However, two problems may make this difficult. First, if the frostbite was associated with hypothermia or lack of proper clothing, the frostbitten area may rewarm as these matters are corrected. Second, the activity of traveling to seek medical care may increase metabolic heat production enough to begin thawing. Even though slow, unintentional rewarming is often unavoidable, refreezing and infection must be avoided at all costs.[102]

If a medical facility cannot be reached promptly, first aid that includes rewarming should be instituted.[109] The patient should be moved into a shelter, to avoid refreezing, and warm water (38-42°C, 100-108°F) prepared. Frostbitten extremities are placed in the water, or the water is poured over a frostbitten nose, ear, or cheek. The injured area should not be rubbed, covered with snow, or placed near a heat source such as an oven. Rewarming frozen skin with water that is hotter than 44°C (111°F) may burn the affected tissues.[109]

The future outcome (prognosis) of frostbite is foreshadowed by various developments during the first 48-72 h. Both good and bad prognostic indicators have been recognized by physicians.[93] These indicators involve damage to skin, loss of sensation (nerve function), and injury to blood vessels (circulation). A good outcome can be expected when watery blisters develop early and sensation, blood flow, and warmth return. A poor prognosis can be expected when no blisters or swelling develop; hard, white, insensitive, cold patches form; dark purple (blood-filled) blisters appear; or physical trauma is superimposed on the frostbite injury.

Currently, the ultimate healing treatment for all frostbite cases does not exist. It also is highly improbable that it will ever exist because the damage is frequently done before arrival at a hospital. This makes prevention the best option. Suggestions for prevention of frostbite appear in the final section of this chapter.

Nonfreezing Injuries Caused by Cold Exposure

Three cold injuries do not involve frozen tissue, but the damage caused by these conditions can be just as debilitating as that caused by frostbite. These conditions are known as trench foot, immersion foot, chilblain, and pernio.

Trench foot and **immersion foot** are so similar that they are often described as the same cold injury.[103] Developing slowly over a period of hours to days, these injuries result from the exposure (>12 h) of wet feet to a temperature of 0.5-10°C (33-50°F). This causes nerve and blood vessel damage, without ice crystallization in cells; this is medically known as a peripheral vasoneuropathy. There are three stages in the progress of immersion foot: (a) a period lasting up to a few days in which the extremity is cold, swollen, numb, and usually without a pulse; (b) a period lasting 2-6 weeks in which blood vessel disturbances, wide temperature gradients, and possible blistering or ulceration appear; and (c) a phase lasting weeks or months in which the patient may be without symptoms, may develop Raynaud's disease (i.e., cold sensitivity involving severe pain), or may have obvious clinical signs and symptoms (itching, pain, edema, numbness, loss of sensation).[4, 103] Trench foot, in contrast to immersion foot, usually occurs on land. Walking, sitting, or standing for lengthy periods wearing wet footgear, coupled with malnutrition and apathy, may lead to a large number of trench foot injuries in soldiers or field laborers.[4]

The appearance of trench foot and immersion foot involves a swollen, cold, irregularly colored extremity. The pulse "bounds" (i.e., is very strong, throbs) and the injured tissue is painful. The skin also may itch or tingle. Loss of a foot or the lower leg is possible.[102]

Chilblain and **pernio** are very similar cold injuries that are less severe than those described above. They are common among youth and women, and are chronically seen in middle-aged females. They affect the legs, toes, hands, and ears most often.[103] Predisposing environmental factors include an air temperature of 0.5-16°C (33-60°F), wind, humidity, and rain. Chilblain results from an exaggerated cold-induced vasoconstriction that results in cell ischemia (localized tissue enemia) and limb edema. In addition, white blood cells infiltrate deep tissue, and blood vessels become inflamed. Thus, chilblain is a vascular disorder with dermatologic manifestations.[108] It is diagnosed by the presence of swollen red areas or plaques, which sometimes progress to blisters and ulcers.[103] This form of cold injury should be treated only by medical personnel.

In cases of pernio, the swelling and itching of chilblain are replaced by superficial burning and pain. Patches of dead cells, often on the hands and feet, eventually slough off, leaving tender tissue beneath. If the depth of the pernio injury were to increase, it eventually would be diagnosed as trench foot.[4] Thus, pernio may be viewed as a condition that lies between trench foot and chilblain.

COUNTERACTING COLD AND WINDCHILL

The two most common and most effective ways for you to prevent cold injury and decrements in exercise performance resulting from cold are (a) limit the duration and severity of exposures to cold air or water and (b) insulate skin surfaces with clothing. Regarding the first option, you certainly should consult a windchill chart (figure 3.2, page 75) to determine the risk of freezing your skin and appendages. You also should be alert for the onset of the clinical signs and symptoms of hypothermia (table 3.4, page 108), frostbite, trench foot, and chilblains. You even may want to consider monitoring body temperature periodically to evaluate heat loss, especially if you are performing low-intensity exercise or if you rest for a prolonged period (table 3.3, page 107).

Regarding clothing, the insulation required to maintain body temperature varies in different environments and with different metabolic rates. The standard unit of clothing insulation is known as the "Clo." One Clo is defined as the insulation necessary to maintain the comfort of a seated adult in a room that has a temperature of 21°C (70°F), 50% rh or less, and air movement of 6 m/min.[20] Street clothing that is comfortable in 21°C (70°F) weather has a total Clo value of approximately 1. Figure 3.15 (top panel) shows three activities (seated clerical work, slow walking, strenuous "power walking") and environments that require insulation of 1 Clo to maintain body temperature. As you can see, an increase of metabolic heat production allows thermal homeostasis at a lower environmental temperature, with 1 Clo of insulation. Obviously, as exercise intensity and ambient windchill decrease, a greater amount of insulation is required. Figure 3.15 (bottom panel) describes the different levels of clothing insulation required for six different metabolic rates (see activities on diagonal lines). The unit of metabolic rate in this figure is known as "Met." It is equivalent to approximately 100 kcal/h in a large person, or 50 kcal/h per square meter of surface area.[20]

Specifically, laborers and athletes should wear multiple layers of light, loose-fitting clothing that insulate the skin with trapped air. An outer garment that provides protection from wind and rain, and allows moisture to escape, is ideal. "Breathable" fabrics such as Gore-Tex™ are ideal for this purpose. Nylon parkas offer protection against severe windchill, but may not offer thermal insulation. Wool and polyester fabrics (usually worn as inner layers) retain some insulative value when wet, but cotton and goose down do not.[110] The layer of fabric that lies in contact with the body ought to transport sweat away from the skin and not retain it. Polypropylene clothing wicks moisture away from the skin, in contrast to cotton fabrics, which retain

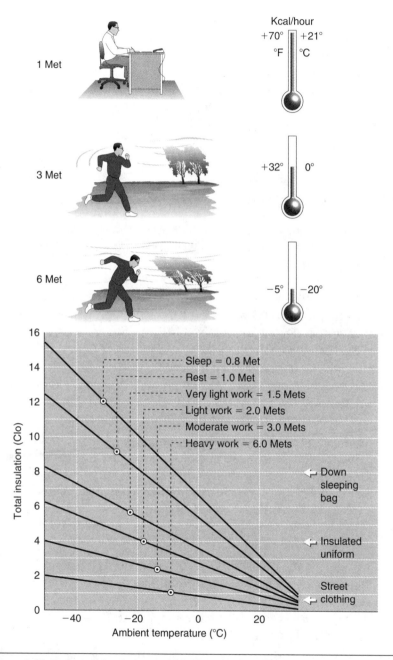

Figure 3.15 Top panel: Insulation provided by 1 Clo, in three selected environments, involving three different activities. Bottom panel: Total insulation of clothing plus trapped air, for comfort at various ambient temperatures and metabolic rates (1 Met equals a metabolic rate of approximately 100 kcal/h).

Top panel adapted from Parsons 1993. Bottom panel adapted from Gonzalez 1988.

moisture. Activity levels also should be considered when selecting the proper layers of clothing. During immobilization or during low-intensity exercise (e.g., bobsled, luge, driving a vehicle), one or two middle layers may be necessary. During high-intensity activities (e.g., running, shoveling snow) the outer layer or head covering may be opened or removed for brief periods to prevent excessive heat storage and sweat accumulation.[111]

Recommendations for Cold Exposure

Numerous other recommendations regarding cold exposure have been proposed and tested by authorities. The following 22 items summarize the most useful of these suggestions.

- Preventing frostbite is much easier than treating it. To prevent frostbite, you should (a) not wear tight clothing, gloves, or boots that restrict circulation; (b) exercise your fingers, toes, and face periodically, to keep them warm and to detect any numb areas; (c) work or exercise with a partner who watches for warning signals of cold injury and hypothermia; and (d) wear the proper insulated clothing and keep it dry.[112] Extra undergarments, socks, and shoes should be carried at all times.

- Because the windchill index (figure 3.2, page 75) contains two components, wind speed and air temperature, it is best to plan for extremely cold weather by dressing to combat the dominant factor. At −23°C (−10°F) and 10 mph wind speed (a windchill of −36°C, −33°F), insulated *coldproof* clothing is in order. At −12°C (+10°F) and 30 mph (also a windchill of −36°C, −33°F), a *windproof* garment should be worn.

- Wet feet greatly increase the risk of cold injury and blisters. A quality sock transfers moisture from your foot to the boot. If two layers of socks are worn (e.g., polypropylene inner layer, wool outer layer), be certain that the toe box of your boot is large enough to allow normal circulation of blood to toes.[113]

- Do not walk through snow in low-cut shoes. If you lack boots or other protective clothing, stay in a sheltered area.[109] Similarly, do not lie or rest directly on snow. Lie on tree boughs, a sleeping pad, or a poncho.

- If you must sleep outdoors, use a sleeping bag but do not wear your clothing. You will sweat, your garments will become damp, and it is likely that you will wake up shivering because your body is hypothermic.

- Skin moisturizing creams should be selected on the basis of their contents. Water-based lotions wet the skin, increasing the likelihood of frostnip or frostbite. Oil-based (i.e., Chapstick, Vaseline) protect the face, hands, and ears from exposure and add virtually no water to the skin.[113]

- Gloves with five fingers are not designed for cold weather. Mittens trap warm air around all fingers and are more effective. Arctic mittens also can be worn with inserts as a second layer, as long as blood flow to fingers is not restricted.[113]

- It is common for athletes to neglect their legs and genitals when dressing for outdoor training. Sweat pants, long underwear, Lycra tights, Gore-Tex pants, or combinations of these garments work well.[114]

- Cold or nearly frozen extremities may be warmed with body heat. Fingers should be placed in the armpit or groin area. Toes and feet also may be placed on the stomach of your travel partner.[113]

- Many cases of frostbite have occurred due to automobile failure in freezing weather. You should keep protective clothing in your vehicle if you travel in isolated areas. When attempting repairs, avoid spilling gasoline on your skin and avoid touching metal with bare hands; always wear gloves. If you are stranded inside the vehicle and must run the engine to warm the inside, use the heater with a window open slightly to guard against carbon monoxide poisoning (see the section titled "Carbon Monoxide" on page 201 in chapter 6).

- Plan your outdoor training course by considering wind direction (i.e., travel into the wind after you have warmed up but before your clothing becomes wet with sweat) and slippery ground. Orthopedic injuries (e.g., muscle strains, ligament or tendon sprains) are the most commonly reported athletic injuries in winter months.[115]

- Prior to exercise or work in the cold, do not wear excessive clothing or exercise to the point of soaking your inner garment with sweat. During exercise, control the pace to limit the rate of sweat production because wet fabrics lose their insulative value. When you conclude a training session in a cold environment, cool down in a sheltered area and change into dry clothing. If clothing remain wet, hypothermia may occur rapidly.

- When you cannot stay warm and dry under the existing conditions of windchill and precipitation, be wise enough to end your exposure and move indoors or into shelter. Be aware of competition, organizational discipline, or peer pressure that might cause you to become overexposed.

- Individuals who have had a previous cold injury are at greater risk of a subsequent cold injury. This may be due to sensitization to cold exposure (i.e., intense vasoconstriction, pain) or behavior that predisposes them to injury. These individuals should be monitored closely.[4]

- Cold exposure affects everyone, but people with Raynaud's disease are especially disabled. This disease involves a localized overreaction to cold via vasoconstriction; a spasm of the smooth muscle in the blood vessels occurs.[116] The circulation is cut off, causing numbness and the nose, fingers, or toes to turn blue. Curiously, these effects may occur when there is a drop in air temperature—even from 21 to 10°C (70 to 50°F). Temperature extremes are not necessary.[117] Athletes or laborers with Raynaud's disease may experience limitations during their runs, skiing, ice-skating, or work shift. Obviously, avoiding situations that produce vasoconstriction is an excellent first step. Activities such as gripping a cold steering wheel or working in a refrigerator should be discouraged.[117] Other methods of dealing with this disease include wearing heavy gloves/mittens, a hat, and insulated shirt; wearing loose clothing; or using electrically warmed mittens. Many patients can be assisted by medications (e.g., nifedipine, reserpine, prazosin) that dilate the blood vessels.[117] Also, a classic conditioning procedure, which can be performed by a patient at home, has proven effective in managing Raynaud's disease. One repeatedly dips one's hands in hot water at the same time that the torso is exposed to cold.[117]

- Dehydration causes you to be less efficient, fatigue earlier, and experience greater physiological strain than when you are normally hydrated. Once dehydrated, however, it is best for you to replace fluids slowly. Rapid replacement of a large volume increases urine production, and cold fluids induce shivering. Research has shown that the best cold weather rehydration technique involves the continuous maintenance of water balance by replacing fluid as it is lost.[110,118] The fluid should be warmed (25-30°C, 77-86°F) and should contain carbohydrate (less than 7%) to delay fatigue and hypoglycemia. Remember that the average adult consumes approximately 2.5 L of fluid each day, in food and beverages; this does not make up for sweat losses. It is likely that consuming 3 L of fluid per day will sustain performance (e.g., walking 16 mi/day) and health during extended living in a cold environment (e.g., 5 days at −23°C [−10°F]).[119]

- Alcohol intoxication increases the risk of cold injury in five ways.[108,111] First, alcohol stimulates skin blood vessel dilation, which increases heat loss. This increased cutaneous blood flow, unfortunately, may give a sense of warmth and well-being, as skin is heated at the expense of the body's core. Second, alcohol inhibits the sensations of cold and pain, thus thwarting normal behavioral responses such as adding clothing or moving indoors. Third, alcohol may inhibit shivering directly, preventing adequate heat production.[108] Fourth, alcohol increases urine production and dehydration. Fifth, it is associated with poor judgment and coordination, resulting in increased accidents and an increased risk of reexposure.

- If you are accidentally immersed in cold water, rest as much as possible to conserve energy. Huddle next to others, if they have fallen into the water.[97] At water temperatures of 20°C (68°F) and below, you are more likely to survive by remaining still. Do not attempt to swim to shore unless it is nearby. Remember that the greatest danger is whole-body hypothermia. Insulate your head, which should be held above the water.

- In life-and-death situations such as cold-water immersion, five personality traits have been identified that enhance survival. These include (a) the will to survive, (b) adaptability and improvisation, (c) optimism/belief that this is only a temporary situation, (d) the ability to tolerate bizarre experiences with calm rationality, and (e) a sense of humor.[120] Be prepared, in any way possible.

- During prolonged living in harsh, cold environments that lasts for weeks and months, the psychological and emotional strains are enormous. The three most potent sources of stress are isolation, monotony, and the absence of hobbies or interest.[121]

- Leaders who are responsible for military, athletic, or industrial groups should provide ample food, replacement clothing, medical care, field warming facilities, evacuation vehicles, and rotation policies. In cold environments, outstanding leaders provide training, discipline, and attention to detail.[4] They should be exposed to the same elements as those whom they manage.

- Advice regarding the management and conduct of large, competitive winter events has been published by the American College of Sports Medicine.[110] This source also provides recommendations for medical personnel, medical aid stations, fluid stations, surveillance of competitors, and educational programs.

REFERENCES

1. Wilson, E. 1972. *Diary of the Terra Nova expedition to the Antarctic.* New York: Humanities Press, 159-160.

2. Berton, P. 1988. *The Arctic grail. The quest for the North West Passage and the North Pole, 1818-1909.* New York: Viking Press, 183.

3. Folk, G.E. 1966. *Introduction to environmental physiology. Environmental extremes and mammalian survival.* Philadelphia: Lea & Febiger, 96-97.

4. Hamlet, M.P. 1988. Human cold injuries. In *Human performance physiology and environmental medicine at terrestrial extremes,* edited by K.B. Pandolf, M.N. Sawka, & R.R. Gonzalez, 435-466. Indianapolis: Benchmark Press.

5. Court, A. 1948. Wind chill. *Bulletin of the American Meteorological Society 29*: 487-493.

6. Milesko-Pytel, D. 1983. Helping the frostbitten patient. *Patient Care 17*: 90-115.

7. Raven, P.B.; Drinkwater, B.L.; Ruhling, R.O.; Bolduan, N.; Taguchi, S.; Fliner, J.; & Horvath, S.M. 1974. Effect of carbon monoxide and peroxyacetyl nitrate on man's maximal aerobic capacity. *Journal of Applied Physiology 36*: 288-293.

8. Regnard, J. 1992. Cold and the airways. *International Journal of Sports Medicine 13*: 5182-5184.

9. O'Cain, C.F.; Dowling, N.B.; Slutsky, A.S.; Hensley, M.J.; Strohl, K.P.; McFadden, E.R.; & Ingram, R.H. 1980. Airway effects of respiratory heat loss in normal subjects. *Journal of Applied Physiology 49*: 875-880.

10. Noakes, T.D. 1986. Body cooling as a method for reducing hyperthermia. *South African Medical Journal 70*: 373-379.

11. Noble, B.J. 1986. *Physiology of exercise and sport.* St. Louis: Times Mirror/ Mosby, 407-423.

12. Raven, P.B.; Nike, I.; Dahms, T.E.; & Horvath S.M. 1970. Compensatory cardiovascular responses during an environmental cold stress, 5°C. *Journal of Applied Physiology 29*: 417-421.

13. Haymes, E.M., & Wells, C.L. 1986. *Environment and human performance.* Champaign, IL: Human Kinetics.

14. Alexander, G. 1979. Cold thermogenesis. In *Environmental physiology III*. Vol. 20, edited by D. Robertshaw, 43-155. Baltimore: University Park Press.

15. Smith, R.M., & Hanna, J.M. 1975. Skinfolds and resting heat loss in cold air and water: Temperature equivalence. *Journal of Applied Physiology 39*: 93-102.

16. Patton, J.F. 1988. The effects of acute cold exposure on exercise performance. *Journal of Applied Sport Science Research 2*: 72-78.

17. Buskirk, E.R.; Thompson, R.H.; & Whedon, G.D. 1963. Metabolic response to cold air in men and women in relation to total body fat content. *Journal of Applied Physiology 18*: 603-612.

18. Toner, M.M., & McArdle, W.D. 1988. Physiological adjustments of man to the cold. In *Human performance physiology and environmental medicine at terrestrial extremes*, edited by K.B. Pandolf, M.N. Sawka, & R.R. Gonzalez, 361-400. Indianapolis: Benchmark Press.

19. Epstein, Y. 1990. Heat intolerance: Predisposing factor or residual injury? *Medicine & Science in Sports & Exercise 22*: 29-35.

20. Gonzalez, R.R. 1988. Biophysics of heat transfer and clothing considerations. In *Human performance physiology and environmental medicine at terrestrial extremes*, edited by K.B. Pandolf, M.N. Sawka, & R.R. Gonzalez, 45-94. Indianapolis: Benchmark Press.

21. Budd, G.M.; Brotherhood, J.R.; Hendrie, A.L; & Jeffery, S.E. 1991. Effects of fitness, fatness, and age on men's responses to whole body cooling in air. *Journal of Applied Physiology 71*: 2387-2393.

22. Falk, B.; Bar-Or, O.; Smolander, J.; & Frost, G. 1994. Response to rest and exercise in the cold: Effects of age and aerobic fitness. *Journal of Applied Physiology 76*: 72-78.

23. Collins, K.J.; Dore, C.; Exton-Smith, A.N.; Fox, R.H.; MacDonald, I.C.; & Woodward, P.M. 1977. Accidental hypothermia and impaired temperature homeostasis in the elderly. *British Medical Journal 1(6057)*: 353-356.

24. Wagner, J.A.; Robinson, R.; & Marins, R.P. 1974. Age and temperature regulation of humans in neutral and cold environments. *Journal of Applied Physiology 37*: 562-565.

25. Armstrong, L.E. 1991. Environmental considerations: Energy expenditure. *National Strength and Conditioning Association Journal 13*: 65-66.

26. Poehlman, E.T.; Gardner, A.W.; & Goran, M.I. 1990. The impact of physical activity and cold exposure on food intake and energy expenditure in man. *Journal of Wilderness Medicine 1*: 265-278.

27. Vallerand, A.L., & Jacobs, I. 1992. Energy metabolism during cold exposure. *International Journal of Sports Medicine 13, Supplement 1*: S191-S193.

28. Gale, E.A.M.; Bennett, T.; Green, H.J.; & MacDonald, I.A. 1981. Hypoglycemia, hypothermia, and shivering in man. *Clinical Science 61*: 463-469.

29. Brooks, G.A.; Fahey, T.D.; & White, T.P. 1996. *Exercise physiology. Human bioenergetics and its applications*. 2nd ed. Mountain View, CA: Mayfield, 438-442.

30. Teitlebaum, A., & Goldman, R.F. 1972. Increased energy cost with multiple clothing layers. *Journal of Applied Physiology 32*: 743-744.

31. Kobrick, J.L., & Fine, B.J. 1983. Climate and human performance. In *The physical environment at work*, edited by D.J. Osborne & M.M. Gruneberg, 69-107. New York: Wiley.

32. Hackney, A.C.; Shaw, J.M.; Hodgdon, J.A.; Coyne, J.T.; & Kelleher, D.L. 1991. Cold exposure during military operations: Effects on anaerobic performance. *Journal of Applied Physiology 71*: 125-130.

33. Doubt, T.J. 1991. Physiology of exercise in the cold. *Sports Medicine (New Zealand) 11*: 367-381.

34. Blomstrand, E.; Bergh, U.; Essen-Gustavsson, B.; & Ekblom, B. 1984. Influence of low muscle temperature on muscle metabolism during intense dynamic exercise. *Acta Physiologica Scandinavica 120*: 229-236.

35. Kruk, B.; Pekkarinen, H.; Hanninen, K.; & Manninen, O. 1991. Comparison in men of physiological responses to exercise of increasing intensity at low and moderate ambient temperatures. *European Journal of Applied Physiology 62*: 353-357.

36. Jacobs, I.; Romet, T.T.; & Kerrigan-Brown, D. 1985. Muscle glycogen depletion during exercise at 9 degrees C and 21 degrees C. *European Journal of Applied Physiology 54*: 35-39.

37. Therminarias, A. 1992. Acute exposure to cold air and metabolic responses to exercise. *International Journal of Sports Medicine 13, Supplement 1*: S187-S190.

38. Timmons, B.A.; Araujo, J.; & Thomas, T.R. 1985. Fat utilization enhanced by exercise in a cold environment. *Medicine & Science in Sports & Exercise 17*: 673-678.

39. Hurley, B.F., & Haymes, E.M. 1982. The effects of rest and exercise in the cold on substrate mobilization and utilization. *Aviation, Space, and Environmental Medicine 53*: 1193-1197.

40. Galloway, S.D.R., & Maughan, R.J. 1997. Effects of ambient temperature on the capacity to perform prolonged cycle exercise in man. *Medicine & Science in Sports & Exercise 29*: 1240-1249.

41. Stromme, S.; Anderson, K.L.; & Eisner, R.W. 1963. Metabolic and thermal responses to muscular exertion in the cold. *Journal of Applied Physiology 18*: 756-763.

42. Anderson, K.L.; Hart, J.S.; Hammel, H.T.; & Sabean, H.B. 1963. Metabolic and thermal response of Eskimos during muscular exertion in the cold. *Journal of Applied Physiology 18*: 613-618.

43. Brebbia, D.R.; Goldman, R.F.; & Buskirk, E.R. 1957. Water vapor loss from the respiratory tract during outdoor exercise in the cold. *Journal of Applied Physiology 11*: 219-222.

44. Hartung, G.H.; Myhre, L.G.; & Nunneley, S.A. 1980. Physiological effects of cold air inhalation during exercise. *Aviation, Space, and Environmental Medicine 51*: 591-594.

45. Bittel, J.H.; Nonotte-Varly, C.; Livecchi-Gonnot, G.H.; Savourey, G.L.; & Hanniquet, A.M. 1988. Physical fitness and thermoregulatory reactions in a cold environment in men. *Journal of Applied Physiology 65*: 1984-1989.

46. Baum, E.; Bruck, K.; & Schwennicke, H.P. 1976. Adaptive modifications in the thermoregulatory system of long-distance runners. *Journal of Applied Physiology 40*: 404-410.

47. Pandolf, K.B., & Young, A.J. 1992. Environmental extremes and endurance performance. In *Endurance in sport*, edited by R.J. Shepard & P.O. Åstrand, 270-282. Oxford, England: Blackwell Scientific.

48. Horvath, S.M. 1981. Exercise in a cold environment. *Exercise and Sport Sciences Reviews* 9:221-263.

49. Noble, B.J. 1986. *Physiology of exercise and sport*. St. Louis: Times Mirror/Mosby, 286-288.

50. Takahashi, H.; Tanaka, M.; Morita, Y.; Igawa, S.; & Kita, H. 1992. Warming-up under cold environments. *Annals of Physiology and Anthropology 11*: 507-516.

51. Rennie, D.W.; Park, Y.; Veicsteinas, A.; & Pendergast, D. 1980. Metabolic and circulatory adaptation to cold water stress. In *Exercise bioenergetics and gas exchange*, edited by P. Cerretelli & B.J. Whipp, 315-321. Amsterdam, Holland: Elsevier/North Holland Biomedical Press.

52. Blake-Reeves, R. 1987. Oxygen affinity. In *Hypoxia and cold*, edited by J.R. Sutton, C.S. Houston, & G. Coates, 89-99. New York: Praeger.

53. Giesbrecht, G.G. 1995. The respiratory system in a cold environment. *Aviation, Space, and Environmental Medicine 66*: 890-902.

54. Davies, C.T.M.; Mecrow, I.K.; & White, M.J. 1982. Contractile properties of the human triceps surae with some observations on the effects of temperature and exercise. *European Journal of Applied Physiology 49*: 255-269.

55. Clarke, R.S.J.; Hellon, R.F.; & Lind, A.R. 1958. The duration of sustained contractions of the human forearm at different muscle temperatures. *Journal of Physiology 143*: 454-473.

56. Bergh, U., & Ekblom, B. 1979. Influence of muscle temperature on maximal muscle strength and power output in human skeletal muscle. *Acta Physiologica Scandinavica 107*: 33-37.

57. Ferretti, G. 1992. Cold and muscle performance. *International Journal of Sports Medicine 13, Supplement 1*: S185-S187.

58. Uksa, J., & Rintamaki, H. 1995. Dynamic work in cold. *Arctic Medical Research 54, Supplement 2*: 29-31.

59. Smith, G.B., & Hames, E.G. 1962. Estimation of tolerance times for cold water immersion. *Aerospace Medicine 33*: 834-840.

60. Nadel, E.R.; Holmer, I.; Bergh, V.; Åstrand, P.O.; & Stolwijk, J.A.J. 1974. Energy exchanges of swimming men. *Journal of Applied Physiology 36*: 465-471.

61. Golden, F.S.; Hampton, I.F.G.; Hervey, G.R.; & Knibbs, A.V. 1979. Shivering

intensity in humans during immersion in cold water. Abstract. *Journal of Physiology (London) 290*: 48P.

62. McCallum, A.L.; McLellan, B.A.; Ross-Reid, S.; & Courtade, W. 1989. Two cases of accidental immersion hypothermia with different outcomes under identical conditions. *Aviation, Space, and Environmental Medicine 60*: 162-165.

63. Keatinge, W. 1965. Death after shipwreck. *British Medical Journal 2*: 1537-1540.

64. Pandolf, K.B.; Toner, M.M.; McArdle, W.D.; Magel, J.R.; & Sawka, M.N. 1987. Influence of body mass, morphology, and gender on thermal responses during immersion in cold water. In *9th International symposium on underwater and hyperbaric physiology*, 145-152. Bethesda, MD: Undersea and Hyperbaric Medical Society.

65. Hayward, J.S.; Eckerson, J.D.; & Collis, M.L. 1975. Thermal balance and survival time prediction of man in cold water. *Canadian Journal of Physiological Pharmacology 53*: 21-32.

66. Hayward, J.S., & Eckerson, J.D. 1984. Physiological responses and survival time prediction for humans in ice water. *Aviation, Space, and Environmental Medicine 55*: 206-212.

67. Toner, M.M.; Sawka, M.N.; Holden, W.L.; & Pandolf, K.B. 1985. Comparison of thermal responses between rest and leg exercise in water. *Journal of Applied Physiology 59*: 248-253.

68. Crittenden, G.; Morlock, J.F.; & Moore, T.O. 1979. Recovery parameters following underwater exercise. *Aerospace Medicine 45*: 1225-1260.

69. Keatinge, W.R. 1961. The effect of work and clothing on the maintenance of body temperature. *Quarterly Journal of Experimental Physiology 46*: 69-82.

70. Martineau, L., & Jacobs, I. 1989. Free fatty acid availability and temperature regulation in cold water. *Journal of Applied Physiology 67*: 2466-2472.

71. Martineau, L., & Jacobs, I. 1989. Muscle glycogen availability and temperature regulation in humans. *Journal of Applied Physiology 66*: 72-78.

72. Neufer, P.D.; Young, A.J.; Sawka, M.N.; & Muza, S.R. 1988. Influence of skeletal muscle glycogen on passive rewarming after hypothermia. *Journal of Applied Physiology 65*: 805-810.

73. Nelson, R.N. 1985. Accidental hypothermia. In *Environmental emergencies*, edited by R.N. Nelson, D.A. Rund, & M.D. Keller, 1-40. Philadelphia: Saunders.

74. Bangs, C.C.; Hamlet, M.P.; & Mills, W.J. 1977. Help for the victim of hypothermia. *Patient Care 12*: 46-50.

75. Giesbrecht, G.G.; Arnett, J.L.; Vela, E.; & Bristow, G.K. 1993. Effect of task complexity on mental performance during immersion hypothermia. *Aviation, Space, and Environmental Medicine 64*: 206-211.

76. Shiraki, K.; Sagawa, S.; Konda, N.; Park, Y.S.; Komatsu, T.; & Hong, S.K. 1986. Energetics of wetsuit diving in Japanese male breath-hold divers. *Journal of Applied Physiology 61*: 1475-1480.

77. Arieli, R.; Kerem, D.; Gonen, A.; Goldenberg, I.; Shoshani, O; Daskalovic, Y.I.;

& Shupak, A. 1997. Thermal status of wet-suited divers using closed circuit O$_2$ apparatus in sea water of 17-18.5 C. *European Journal of Applied Physiology 76*: 69-74.

78. Allan, J.R. 1988. A technical basis for the development of thermal performance standards for immersion protection. In *Environmental ergonomics*, edited by I.B. Mekjavic, E.W. Banister, & J.B. Morrison, 205-220. Philadelphia: Taylor and Francis.

79. Frisancho, A.R. 1993. *Human adaptation and accommodation*. Ann Arbor, MI: The University of Michigan Press, 101-143.

80. Adams, T., & Covino, B.G. 1958. Racial variations to a standardized cold stress. *Journal of Applied Physiology 12*: 9-12.

81. Hammel, H.T.; Elsner, R.W.; LeMessurier, D.H.; Anderson, H.T.; & Milan, F.A. 1959. Thermal and metabolic responses of the Australian Aborigine exposed to moderate cold in summer. *Journal of Applied Physiology 14*: 605-615.

82. Young, A.J. 1996. Homeostatic responses to prolonged cold exposure: Human cold acclimatization. In *Handbook of physiology*. Section 4: *Environmental physiology*. Vol. 1, edited by M.J. Fregly & C.M. Blatteis, 419-438. New York: Oxford University Press.

83. Bittel, J.H.M. 1987. Heat debt as an index for cold adaptation in men. *Journal of Applied Physiology 62*: 1627-1634.

84. LeBlanc, J. 1962. Local adaptation to cold of Gaspe' fisherman. *Journal of Applied Physiology 17*: 950-952.

85. Skreslet, S., & Aarefjord, F. 1968. Acclimatization to cold in man induced by frequent scuba diving in cold water. *Journal of Applied Physiology 24*: 177-181.

86. Mathew, L.; Purkayastha, S.S.; Jayashankar, A.; & Nayar, H.S. 1981. *Study on cold acclimatization in men*. Delhi, India: Defense Institute of Physiology and Allied Sciences, 1-10.

87. Bruck, K. 1976. Cold adaptation in man. In *Regulation of depressed metabolism and thermogenesis*, edited by L. Jansky & X.J. Mussachia, 46. Springfield, IL: Charles C Thomas.

88. Shephard, R.J. 1993. Metabolic adaptation to exercise in the cold. *Sports Medicine 16*: 266-289.

89. Petrofsky, J.S.; Burse, R.L.; & Lind, A.R. 1981. The effect of deep muscle temperature on the cardiovascular responses of man to static effort. *European Journal of Applied Physiology and Occupational Physiology 47*: 7-16.

90. Shephard, R.J. 1992. Fat metabolism, exercise, and the cold. *Canadian Journal of Sport Sciences 17*: 83-90.

91. Bruck, K. 1980. Basic mechanisms in longtime thermal adaptation. In *Advances in physiological sciences*. Vol. 23, edited by Z. Szelenyi & M. Szekely, 263. Oxford, England: Pergamon Press.

92. Blue, B. 1994. Safe exercise in the cold and cold injuries. In *Sports medicine secrets*, edited by M.B. Mellion, 86-90. Philadelphia: Hanley and Belfus.

93. Departments of the Army, Navy, Air Force. 1985. *Cold injury* (Technical Bulletin 81). St. Louis: U.S. Army Adjutant General Publications Center, 1-14.

94. Bangs, C., & Hamlet, M. 1980. Out in the cold—Management of hypothermia, immersion, and frostbite. *Topics in Emergency Medicine 2*: 19-37.

95. Danzl, D.F.; Pozos, R.S.; & Hamlet, M.P. 1995. Accidental hypothermia. In *Wilderness medicine. Management of wilderness and environmental emergencies,* 3rd ed., edited by P.S. Auerbach, 51-103. St. Louis: Mosby.

96. Burr, L. 1983. Accidental hypothermia: Always a danger. *Patient Care 17*: 116-153.

97. Fears, J.W. 1986. *Complete book of outdoor survival.* New York: Outdoor Life Books, 273-280.

98. Pugh, L.G.C. 1964. Deaths from exposure on Four Inns Walking Competition, March 14-15. *Lancet i*: 1210-1212.

99. Pugh, L.G.C. 1967. Cold stress and muscular exercise with special reference to accidental hypothermia. *British Medical Journal 2*: 333-337.

100. Parsons, K.C. 1993. Cold stress. In *Human thermal environments*, 181-198. London: Taylor and Francis.

101. Hackett, P.; Mills, W.; & Nemeroff, M. 1989. *State of Alaska hypothermia and near drowning guidelines.* Juneau, AK: Alaska Department of Health and Social Services, Emergency Medical Services.

102. Schimelpfenig, T., & Lindsey, L. 1991. Cold injuries. In *NOLS wilderness first aid*, 153-179. Lander, WY: NOLS Publications.

103. McCauley, R.L.; Smith, D.J.; Robson, M.C.; & Heggers, J.P. 1995. Frostbite and other cold-induced injuries. In *Wilderness medicine. Management of wilderness and environmental emergencies,* 3rd ed., edited by P.S. Auerbach, 129-152. St. Louis: Mosby.

104. Foray, J. 1992. Mountain frostbite. *International Journal of Sports Medicine 13*: S193-S196.

105. Robson, M.C., & Heggers, J.P. 1981. Evaluation of hand frostbite blister fluid as a clue to pathogenesis. *Journal of Hand Surgery 6*: 43-47.

106. Reus, W.F. 1984. Acute effects of tobacco smoking on blood flow in the cutaneous micro-circulation. *British Journal of Plastic Surgery 37*: 213-219.

107. Bangs, C.C. 1984. Cold injuries. In *Sports medicine*, edited by R.H. Strauss, 323-343. Philadelphia: Saunders.

108. Nelson, R.N. 1985. Peripheral cold injury. In *Environmental emergencies*, edited by R.N. Nelson, D.A. Rund, & M.D. Keller, 25-40. Philadelphia: Saunders.

109. Milesko-Pytel, D. 1983. Helping the frostbitten patient. *Patient Care 17*: 90-115.

110. Armstrong, L.E.; Epstein, Y.; Greenleaf, J.E.; Haymes, E.M.; Hubbard, R.W.; Roberts, W.O.; & Thompson, P.D. 1996. American College of Sports Medicine position stand. Heat and cold illnesses during distance running. *Medicine & Science in Sports & Exercise 28(12)*: i-x.

111. Armstrong, L.E., & Hamlet, M.P. 1989. Preventing cold injuries during training, competition, and recreation. *National Strength and Conditioning Association Journal 11*: 51-53.

112. Goulart, B. 1986. Dressed to chill. *Soldiers 1*: 35-36.

113. U.S. Army Rangers. 1988. *Prevention of cold injury*. Fort Benning, GA: Ranger Training Brigade, 1-30.

114. Stamford, B. 1995. Smart dressing for cold weather workouts. *The Physician and Sportsmedicine 23*: 105-106.

115. Armstrong, L.E. 1992. The most common athletic injuries during winter months. *National Strength and Conditioning Association Journal 14*: 85.

116. Weiss, H. 1992. *Secrets of warmth*. Seattle: Cloudcap, 1-158.

117. Hamlet, M.P. 1990. Raynaud's disease: A simple approach to management. *The Physician and Sportsmedicine 18*: 129-132.

118. Rintamaki, H.; Makinen, I.; Uksa, J.; & Latvala, J. 1995. Water balance and physical performance in cold. *Arctic Medical Research 54, Supplement 2*: 32-36.

119. Roberts, D.E.; Patton, J.F.; Pennycook, J.W.; Jacey, M.J.; Tappan, D.V.; Gray, P.; & Heyder, E. 1984. *Effects of restricted water intake on performance in a cold environment* (Technical Report T2-84). Groton, CT: Naval Submarine Medical Research Laboratory, 1-19.

120. Jenkins, M. 1997. The test of your life. *Men's Health 12*: 160-164.

121. Anonymous. 1993. Mind and body over ice. *The Independent (London)*, 16 February, 10.

chapter 4

Diving into Earth's Hyperbaric Environment

The swimmer may prepare for diving by taking a slow and full inspiration, letting himself sink gently into the water, and expelling the breath by degrees, when the heart begins to beat strongly.

In order to descend in diving, the head must be bent forward upon the breast; the back must be made round; and the legs must be thrown out with greater vigor than usual; but the arms and hands, instead of being struck forward as in swimming, must move rather backward, or come out lower, and pass more behind....

Still in swimming between top and bottom, the head must be kept a little downward, and the feet must be thrown out a little higher than when swimming on the surface; and if the swimmer thinks that he approaches too near the surface, he must press the palms upward.

To ascend the chin must be held up, the back made concave, the hands stuck out high, and brought briskly down.

—D. Walker, *British Manly Exercises. Containing Rowing and Sailing, Riding and Diving, etc.,* 1836[1]

Diving disrupts the body's **homeostasis** of internal pressure and gas concentrations. Both involve changes in the composition of, and movement of, body fluids. The cardiovascular and respiratory systems are essential to the regulation of pressure and gases. However, when the excessive pressure of deepwater diving (hyperbaria) overwhelms the adaptive capacity of these two systems, diving illnesses and injuries occur.

Although your body can acclimatize in minor ways to hyperbaria, specialized equipment is required to assist normal physiological responses and provide gases to the lungs. Breathing precise mixtures of pressurized gases, for example, requires considerable technical expertise. Novice divers and females should take special precautions. In this chapter, these and other safety recommendations are presented to help make sport diving a safe, enjoyable endeavor.

THE HISTORY OF DIVING

The excerpt that opens this chapter is believed to be the first description of breath-hold diving as a sporting activity (figure 4.1). Yet, commercial diving has existed since at least 460 B.C. The accounts of the Greek historian Herodotus explain that the diver Scyllis was employed by the Persian king Xerxes to recover treasure from ships wrecked in the Mediterranean Sea during the 50-year war between Greece and Persia.[2, 3] Similar accounts of military and salvage operations appear throughout the histories of the Roman and Japanese cultures.[3, 4] Military divers cut anchor cables to set enemy ships adrift, bored or punched holes in the bottom of ships, and built harbor defenses.[5]

However, human underwater exploits were limited to breath-hold diving until about three hundred to four hundred years ago, when a series of technological developments prolonged submergence.[3] These advances provided external air to divers in a variety of ways. In 1531, for example, a practical diving bell was produced. Literally bell-shaped, with the bottom open to the sea, this device was weighted

Figure 4.1 Walker's 1836 depiction of breath-hold diving as a sport.[1]

and sank in a vertical position. A diver could either remain inside the bell, if positioned directly over his work by the support ship, or could venture outside for short periods of breath-holding activity.[5] Much later, in 1690, astronomer Sir Edmund Halley devised a leather tube to carry surface air to submerged barrels; this tube supplied air to manned diving bells at a depth of 18 m (60 ft).[2] In an early demonstration of this system, Halley remained at 18 m under the Thames River in Great Britain, with four other people, for almost 1.5 h.[5]

In 1715, Englishman John Lethbridge developed the first one-person diving suit (figure 4.2, page 134). His apparatus was basically a reinforced, leather-covered barrel of trapped air, equipped with a glass window and two armholes with watertight sleeves, allowing useful work for up to 34 min.[5] This apparatus and succeeding equipment suffered from the same limitations as diving bells. They contained no practical way to supply air continuously to a diver.

Fortunately, during a 90-year period beginning in 1788, four vital technological breakthroughs enhanced diving. The first was a pump that delivered compressed air from the surface to either a diving bell or a diver.[6] The second was a full-length waterproof diving suit, invented by August Siebe in 1840, that included a helmet with viewing ports, surface-supplied air, and an exhaust valve (figure 4.3).[5] The third breakthrough, the "demand regulator," was patented by a French engineer in 1866.[5] This device supplied air to divers, on the demand of inhaling,

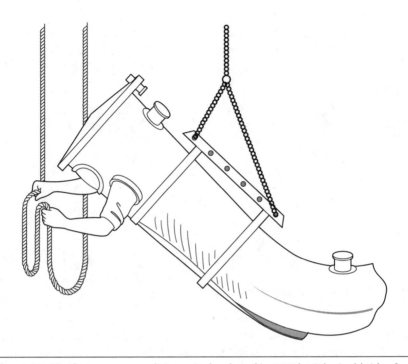

Figure 4.2 The first one-person diving suit, developed in 1715 by John Lethbridge.[5]

at pressures that were much greater than the 1 atmosphere of pressure (1 atm) found at sea level.[3] The final advance was the first successful self-contained breathing apparatus. Developed in 1878 by Henry Fluess and August Siebe, this apparatus utilized pure oxygen with a carbon dioxide absorbent system (see next paragraph). Although it exposed divers to the risk of oxygen poisoning (the breathing of gas containing O_2 at very high levels), no cases of this illness were recorded during extended underwater operations in a flooded tunnel.[2]

The next major improvement in underwater equipment did not arise until 1943, as a part of the French resistance activities against Nazi Germany. Renowned undersea explorer Jacques-Yves Cousteau and Emile Gagnon combined a demand regulator with a compressed air tank, forming what they called a **s**elf-**c**ontained **u**nderwater **b**reathing **a**pparatus (scuba),[5] the first truly efficient and safe open-circuit device of its kind. In an *open-circuit* design, pressurized air is taken from a supply tank, inhaled, and exhausted to the surrounding water. But, because this design was detrimental to military operations since it left a telltale trail of bubbles on the water's surface, and because it

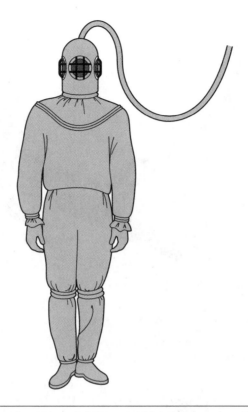

Figure 4.3 Siebe's full-length waterproof diving suit with helmet, produced in 1840.[5]

wasted compressed air, various *closed-circuit* breathing apparatuses were soon developed that eliminated these two disadvantages. When using a closed-circuit scuba system, a diver breathes either pure O_2 (which limits depth because of potential oxygen poisoning) or a gas mixture containing oxygen and an inert gas (e.g., helium, nitrogen), and the expired gas is recirculated, not released to the surrounding water. A chemical filter removes carbon dioxide while oxygen is added slowly from the tank, to replace that which is consumed by the body. Today's closed-circuit scuba equipment avoids oxygen poisoning by electronically sensing O_2 and providing a constant concentration at any depth[5] (see section below, "Medical Considerations: Breathing Gas Mixtures," page 149).

After World War II, the development and sale of open-circuit scuba equipment (figure 4.4) to the general public made the underwater world accessible to increasing numbers of people. The growth of

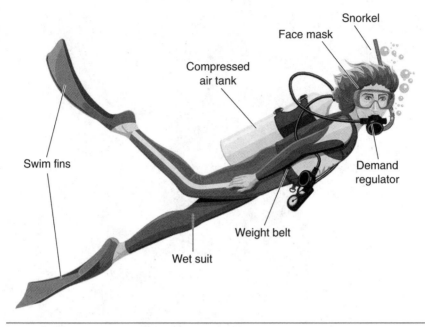

Figure 4.4 Modern scuba diving equipment.
Adapted from Graver 1993.

diving as a sport has been especially noteworthy. Since scuba was first introduced to the United States, over 7 million Americans have been certified as recreational or sport divers.[3] Internationally, commercial divers have opened an entire industry—offshore petroleum production. Further, geologists, biologists, zoologists, and archaeologists have gone underwater to seek new clues to the origin and behavior of the earth and its many life forms.

YOUR BODY UNDER PRESSURE

The following paragraphs explore the physiological reasons why each of the above technical advances was necessary. Numerous medical problems may arise when humans dive to great depths because of the tremendous pressure that water exerts on the tissues and organs of the body. These problems occur despite the safety features found in modern scuba gear. The causes and treatments of these medical conditions are described at the end of this chapter, as they relate to water pressure. The effects of cold-water immersion were described in chapter 3.

Diving, Pressure, and Gas Volumes

As you read these words, you are under a very tall column of air that has weight but cannot be sensed by your central nervous system. The measurable weight of this vertical column of atmospheric gases is known as **air pressure**. When you are immersed, water exerts a much greater pressure than air, **hydrostatic pressure,** from all directions toward the center of your body.

The three states of matter—solid, liquid, and gas—have differing degrees of compressibility. Although solids and liquids are relatively incompressible, gases expand and contract readily in proportion to pressure and temperature changes. Because the body consists mostly of water and other incompressible materials, as a scuba diver descends into deep water the increased ambient pressure is evenly transmitted throughout the organs and tissues. Because continuous gas transport (e.g., O_2, CO_2) is essential to metabolism, health, and performance, it is important that you understand the effects that diving has on the gases in the lungs, bloodstream, and tissues. This also will provide you with a better understanding of the clinical problems encountered by divers.

Depth and Hydrostatic Pressure

Figure 4.5 (page 138) depicts the effect of water depth on the volume of gas inside a closed container. On the surface, air presses on the gases inside the vessel with a force of 1 atm. An underwater environment is the only place on earth where pressure on the body greatly exceeds 1 atm; this zone is often called a **hyperbaric** (high pressure) environment. At a depth of 10 m (33 ft), the external pressure is equivalent to 2 atm. Lowering the container from sea level to a depth of 20 m (66 ft) produces an absolute ambient pressure of 3 atm. At 30 m (99 ft), the hydrostatic pressure is 4 atm, and so on. Because the volume of a gas is inversely proportional to the external pressure applied to it (this principle is known as Boyle's law), the volume of air trapped in the closed container is halved at a depth of 10 m, and is reduced to one-third and one-fourth of the original volume at depths of 20 m and 30 m, respectively.[7]

However, human lungs are not as rigid as a metal vessel and, as a diver descends, fluids and gases move between compartments in the body. Further, enclosure in a pressurized vessel (e.g., a submarine), or breathing pressurized gas, alters the effect of hydrostatic pressure on lung gas volume. Figure 4.6 illustrates how your lung gases

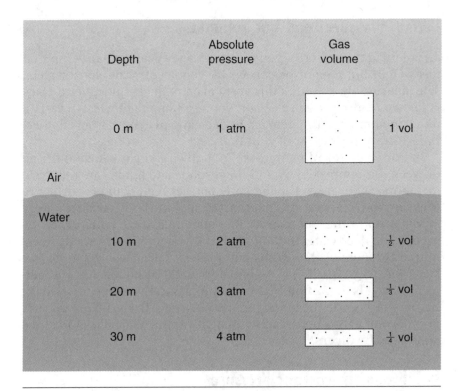

Figure 4.5 At greater depths, pressure increases and the volume of a gas decreases, in proportion to the change in pressure.
Adapted from Strauss 1984.

would be affected if you descended to a depth of 30 m, using three different modes of underwater exploration: a submarine, breath-hold diving, or breathing compressed gas from scuba tanks.[8] The hull of a submarine is designed to withstand immense water pressure, while the air pressure inside the vessel is maintained at 1 atm. Therefore, passengers inside a *submarine* experience no change in lung pressure or volume during travel at a depth of 30 m or more. On the other hand, the lungs are compressed during *breath-hold diving* and the gas pressure within them is approximately the same as the surrounding hydrostatic pressure (4 atm at a depth of 30 m). This causes the gas inside the lungs to compress to one-fourth of its original volume. While wearing *scuba gear*, the gas pressure within the lungs remains about that of the external water pressure because a pressurized gas mixture is provided at 4 atm through the demand regulator, and the diver breathes normally. Because of the relative hyperbaric state inside the lungs, the density (i.e., the number of particles per cubic centi-

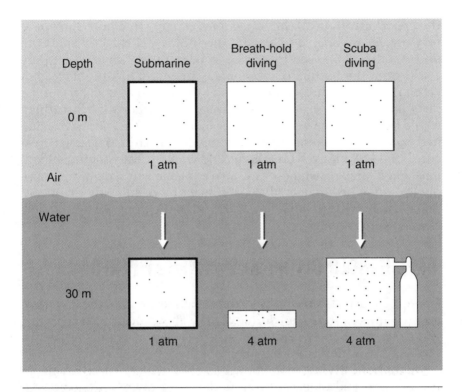

Figure 4.6 Effects on lungs of a descent to a depth of 30 m, using three different modes of underwater exploration.
Adapted from Strauss 1984.

meter) of the compressed gas mixture is greater than that found at sea level. As described below, these properties of compressed gases require that specific techniques be used during ascent (see section titled "Medical Considerations: Breathing Gas Mixtures," page 149).

Snorkeling, a variety of breath-hold diving, is not depicted in figure 4.6 because it typically is performed in shallow water only. The snorkel tube is 1.3-1.9 cm (0.5-0.75 in.) in diameter, is fitted with a mouthpiece at one end, and is open to the air at the other end; this allows the diver to swim in a prone position. As the simplest of underwater breathing devices, a snorkel is an artificial extension of the diver's own airways. As such, the diver's lungs and the snorkel tube communicate with the air and remain at 1 atm of pressure, regardless of the depth to which the diver descends.[8] The deeper the descent, the greater the imbalance of pressure on the lungs and the more difficult it is to breathe (e.g., 1 atm internal versus 2 atm external pressure at a depth of 10 m). In fact, at a depth of only 0.9 m (3 ft), the hyperbaric force of water becomes so large that human respiratory organs are unable to expand the thorax.[7] This fact dispels the notion, often expressed by novice swimmers and divers, that they could sit at the bottom of a swimming pool and breathe through a garden hose.[7] If a snorkel were much longer than 30 cm (1 ft), or if a diver attempted to use a long breathing tube underwater, breathing would be impossible because (a) the inspiratory muscles have a limited capacity to overcome great differences in external-internal pressure,[9] and (b) a longer breathing tube would create additional volume and airway resistance. This also explains why a diver's air supply line must be connected to a surface pump, and why the invention of such a device in 1788 was critical to the progress of commercial and sport diving.

BREATH-HOLD DIVING AND PHYSICAL PERFORMANCE

Figure 4.7 illustrates the body systems that respond to the underwater pressures experienced during diving. These include the nervous, endocrine, respiratory, cardiovascular, muscular, and urinary systems. The text below describes the nature of these responses.

Breath-hold diving, also known as skin diving or free diving, is the simplest and oldest form of underwater diving. Archeological relics, dating back to 4500 B.C., include objects crafted from pearl shells that must have been obtained by breath-hold divers.[10] The most widely studied group of breath-hold divers are the ama of Japan and Korea.[11] The word **ama** means "sea-woman" in Japanese, but this term is accepted for both males and females. The foundational investigations of ama divers were conducted by Teruoka in 1932.[4, 11] He published detailed accounts of their underwater operations, diving patterns, equipment, seasonal variations, and respiratory gas exchange.[12]

The work shift of ama divers lasts approximately twice as long in summer as in winter.[13] Wearing a traditional cotton suit results in a net heat loss of about 10 kcal/min in summer and about 30 kcal/min in winter.[14] This difference demonstrates the important role that water temperature plays in the duration of underwater labor. As described in chapter 3 (see section titled "Cold-Water Immersion and Physical Performance," page 94), water-immersion hypothermia limits exercise performance. Ama cease activity when core body temperature reaches 95°F (35°C), but they resume diving for one or more shifts following rewarming, on a boat or on land.[14] Wet-suited divers find hypothermia to be less of a problem due to the insulation provided.

The Body's Responses to Stress

Nervous System

Maintains homeostasis:

- hypothalamus receives afferent input
- sends efferent messages in response to changes in blood gas levels or skin/core temperature

Endocrine System

- fluid-electrolyte hormones conserve water and NaCl via the kidneys

Respiratory System

- hyperbaric acclimatization results in enhanced sensitivity to changes in blood gases (chemoreceptors) and an increase in lung vital capacity

Cardiovascular System

- cutaneous blood vessels constrict or dilate, depending on water temperature
- pressure and volume receptors respond to underwater depths by altering blood pressure, heart rate, and cardiac output
- the incidence of cardiac arrhythmias increase

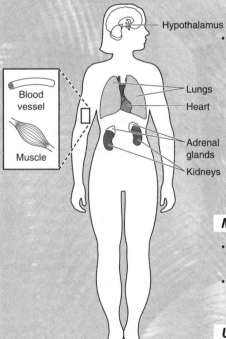

Hypothalamus

Blood vessel

Muscle

Lungs

Heart

Adrenal glands

Kidneys

Muscular System

- in cold water, shivering produces internal metabolic heat
- respiratory muscles adapt to hyperbaria

Urinary System

- kidneys are involved in fluid-electrolyte homeostasis

Figure 4.7 Systems of the body that respond to underwater hyperbaria and changes in blood O_2 and CO_2 levels.

The cardiovascular demands of diving are submaximal, but occasional brief, strenuous bursts of activity are required. Maximal or near-maximal exercise is rarely performed underwater, due to the risks and respiratory limitations inherent in both breath-hold and scuba diving, and because of the required waiting periods during ascent from great depths. In fact, the major limitations on underwater performance are often viewed as primarily psychological or perceptual, not physiological, factors.[15] However, the 1996 review by Lin and Hong[11] provides examples of the limits of human performance during breath-hold diving. They summarized the characteristics of summer work shifts (range of mean values, 1989 to 1991) for male and female ama during unassisted dives as follows:

- work-shift duration, 170-201 min
- number of dives, 109-115 per work shift
- average depth of each dive, 3.7-6.7 m (12-22 ft)
- work-shift surface time, 113-138 min
- work-shift diving time, 52-63 min
- single-dive descent time, 7-10 sec
- single-dive bottom time, 13-18 sec
- single-dive total time, 27-37 sec

These data indicate that the duration of a single commercial dive is usually less than 60 sec. In contrast, other situations may make it possible to stay underwater for a considerably longer period. For example, underwater breath-hold times of 4.0 and 4.5 min have been observed in competition and during research studies,[16] when **hyperventilation** (i.e., repetitive, rapid, and deep air breathing) was used prior to the effort. The current world record for underwater breath holding in a swimming pool is 7 min 35 sec for men (Andy Le Sance; La Reunion, France; 1995) and 5 min 41 sec for women (Corinne Delomenede; La Reunion, France; 1998). And, when air hyperventilation was followed by three deep breaths of pure O_2, two laboratory subjects were able to refrain from breathing for 15.2 and 20.1 min, respectively.[17]

MEDICAL CONSIDERATIONS: BREATH-HOLD DIVING

As a specialized skill, breath-hold is not without certain risks. Risks associated with breath-hold diving include complications related to predive hyperventilation and barotrauma.

Predive Hyperventilation: Useful but Risky

Although hyperventilation is the most effective means of prolonging underwater breath-hold times,[11] it increases the risk of drowning, due to the following factors:[7, 18]

1. The level of carbon dioxide (CO_2) in the blood is greatly reduced and the diver's urge to breathe is markedly reduced or absent.

2. As the diver descends, hydrostatic pressure compresses the thorax and maintains O_2 concentration at a relatively high level (see "Depth and Hydrostatic Pressure," page 137). Even though O_2 leaves the lung and attaches to hemoglobin in the blood, an adequate concentration of O_2 is maintained because of the increased water pressure. Thus, low oxygen levels do not motivate the diver to surface and breathe again.

3. The diver stays on the bottom and consumes O_2 until blood CO_2 rises to a critical level. At this point, there is barely enough O_2 in the body to allow exercise to continue, and the diver may suddenly lose consciousness (black out) before reaching the surface.

© Jim West

Hyperbaric environments expose divers to a variety of physiological stressors not found at the surface.

Approximately 7000 deaths occur by drowning in the United States each year.[18] In an attempt to identify possible causes, Albert Craig from the University of Rochester reconstructed the events that preceded near-drowning accidents.[6, 19] He learned that all survivors hyperventilated before going underwater and also that the swimmers usually felt the urge to breathe, but had little or no warning that they were going to pass out.[18] Most of the survivors also stated that they had some goal in mind or were in competition with others;[19] this included work-related tasks, such as ama divers collecting valuable abalone shells. Based on this information, Craig concluded that **breath-hold diving blackout** results from hypoxia (low blood O_2 concentration), as described in the previous paragraph.

Hyperbaric Injuries

It is possible to damage the lungs during breath-hold dives that reach depths of 30 m (100 ft) or more. Persons with the unusual syndrome known as **lung squeeze**, experience shortness of breath, cough up frothy blood, and exhibit pulmonary edema (fluid in the lungs) on a chest X ray.[3] Medical treatment involves supplemental oxygen and

© Mark Collins/www.norbertwu.com

Careful monitoring of scuba equipment is essential for safe dives.

respiratory support as needed. The symptoms usually resolve within a few days without further complications.

Lung squeeze is medically classified as a form of **barotrauma** (*baro*, "pressure"; *trauma*, "injury") because of the role that hydrostatic pressure plays in its development. When a diver descends to a depth at which total lung volume becomes less than normal residual volume (i.e., the small amount of air that remains in the lungs even after the most forceful expiration; approximately 1.0-1.5 L),[5] insufficient air volume exists in the lung to balance pressure across the lung tissue. Thus, the internal pressure (from blood to alveoli) exceeds the pressure within the alveoli. This results in a movement of fluid into the lung and greatly reduces O_2 diffusion across the pleural membrane, from the lung to the blood.[3] If **alveolar rupture** occurs, blood is literally sucked into the airways and a diver may drown in his own body fluids.[7]

Often, when lung tissue bursts and air is forced through the alveoli, the air migrates laterally to burst through the sac covering the outer surface of the lungs. This causes a bolus of air to form in the chest cavity, between the inner chest wall and the outer lung tissue. If this air pocket continues to expand, it causes the lung to collapse and pushes the heart and other internal organs aside. Known as **pneumothorax**, this condition requires surgery, using a syringe to extract the air bubble.[7]

There is considerable variability among individuals in the safe depth for breath-hold diving without danger of lung squeeze. There is no danger if sufficient air volume is present in the lung to balance pressure changes that occur during descent. For most people, this critical depth is about 30 m (100 ft).[7] The current world record for the deepest breath-hold dive is 135 m (443 ft), held by Gianluca Genoni (Sardinia; Italy, 1998). Tanya Streeter performed the deepest breath-hold dive by a woman (113 m [370 ft]), at Cabo San Lucas, Mexico, in 1998. During descent, they both carried weights and then inflated balloons during ascent.

Divers may experience other forms of barotrauma if the gas pressure in the internal cavities does not equalize with the external hydrostatic pressure. Figure 4.8 (page 146) illustrates the most common of these, **middle-ear squeeze** (i.e., barotitis media, or middle-ear barotrauma). This disorder, which affects more than 40% of all breath-hold and scuba divers, is a direct result of Boyle's law and the anatomy of the eustachian tube.[3,20] You may have experienced the pain of middle-ear squeeze as you descended to the bottom of a deep pool while holding your breath. The top panel of figure 4.8 shows a normal ear at the water's surface, with pressure equalized on both sides of the tympanic membrane at 1 atm.[3] The bottom panel depicts the diver's ear at a depth of 1.2 m (4 ft). Middle-ear squeeze occurs because the eustachian tube is closed (due to a mucus plug or inflammation), the pressure in the middle ear cannot equilibrate with air in the lungs via the trachea, and the external water pressure of 1.3 atm is greater than the 1 atm of pressure in the middle ear. This condition may be painful in water as shallow as 0.6-0.9 m (2-3 ft). If a diver does not heed the symptoms of middle-ear squeeze and allows the pressure difference to reach 1.1-1.5 atm (at depths of 1.3-5.3 m, 4.3-17.4 ft), the pressure imbalance may be resolved unsatisfactorily by rupture of the tympanic membrane; occasionally patients will have a small amount of blood around the nose or mouth and a mild hearing loss.[3] Thus, earplugs should never be worn while diving, because

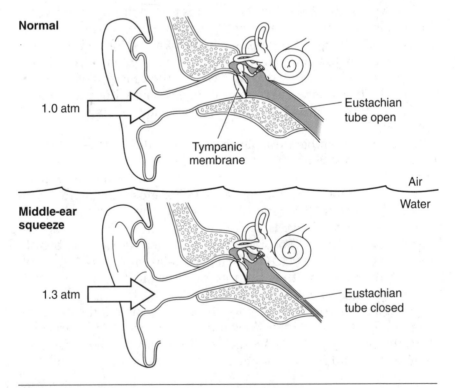

Figure 4.8 Middle-ear squeeze. Top panel: A normal ear at sea level (1 atm), with pressure equalized on both sides of the tympanic membrane. Bottom panel: A normal ear at a depth of 1.2 m (4 ft). Because the eustachian tube is closed, the middle-ear pressure cannot equalize with external air pressure, and painful symptoms result. Adapted from Kizer 1995.

hydrostatic pressure pushes the earplug deep into the external ear canal; this makes it impossible for water to enter the external ear canal, and equalize pressure on both sides of the tympanic membrane. If a pocket of surface air gets trapped between the plug and the eardrum, the eardrum may rupture during descent.[7] Prevention is the key to avoiding middle-ear squeeze. Experienced divers understand this disorder and inflate the middle ear immediately on submerging. This is accomplished safely via the Frenzel maneuver, either with the mouth open or with a mouthpiece and demand regulator in place. The diver increases pressure within the pharynx by occluding the nose, closing the glottis, and contracting the pharynx muscles. Air is thus forced through the eustachian tubes without altering thoracic pressure.[8]

Several other forms of barotrauma occur when the gases in a body cavity compress or expand during descent or ascent. These conditions may occur during breath-hold diving, scuba diving, or both. The most common hyperbaric diving injuries are summarized below.[3, 5, 7]

- **Sinus squeeze** has essentially the same cause as middle-ear squeeze (i.e., increased pressure during descent) but is less common. If the pressure in any nasal sinus cannot be equalized with the surrounding water pressure, a relative vacuum develops in the sinus cavity. Intense pain and bleeding follow the barotrauma to the sinus membrane. This condition usually occurs in a diver who has an upper respiratory infection or allergies.

- **Suit squeeze** and **external-ear squeeze** occur when skin (or an ear) becomes trapped beneath a fold or wrinkle of a diver's dry suit, causing a closed airspace. During descent, the pressure-induced contraction of air under the suit results in a partial vacuum and a dramatic pinching/reddening of the skin (ecchymosis). This condition requires no treatment and resolves within a few days, unless the eardrum ruptures.

- **Face-mask squeeze** is caused by a failure to equalize pressure during descent, via nasal exhalation, inside a partial mask that covers the eyes and nose. With a full face mask, this condition may be caused by a faulty air supply or malfunction of a pressure valve. This condition may result in ruptured blood vessels in the eye or, in severe cases, dislocation of the eyes from the orbit of the skull.

- **Gastrointestinal barotrauma** involves fullness, belching, flatulence, and abdominal pain due to bowel gas that expands during ascent. It is most often observed in novice divers and those who chew gum during a dive, drink carbonated beverages, or eat legumes prior to diving. This condition is usually remedied by elimination of the excess gas.

• **Alternobaric vertigo (ABV)** usually occurs during ascent and is due to the sudden development of unequal middle-ear pressure, which causes pronounced vertigo (the sensation that one's surroundings are swirling). ABV also may precipitate a panic response, leading to near drowning.

• **Air embolism** involves the expansion of lung gases during ascent. If, for example, a diver takes a full breath of air while at a depth of 30 m (100 ft, water pressure of 4 atm) and begins ascent *without exhaling*, the volume of this air will expand four times by the time the diver reaches the surface (see Boyle's law in the previous section titled "Depth and Hydrostatic Pressure," page 137). Because the lung cannot hold such a large volume of gas, the diver must allow this trapped air to escape, or suffer the consequences. Specifically, alveoli rupture and air bubbles are forced across the pulmonary capillaries into the blood. These bubbles enter the left atrium, the left

© Norbert Wu/www.norbertwu.com

Although the underwater world often seems serene and peaceful, caution must be practiced to avoid hyperbaric injuries.

ventricle, the aorta, and eventually the coronary and cerebral arteries. The results are tragic, in that a heart attack or stroke usually occurs. Fatal air embolism has occurred from breath-holding during ascent from a dive in only 1.8 m (6 ft) of water.[3]

MEDICAL CONSIDERATIONS: BREATHING GAS MIXTURES

Air consists of approximately 21% O_2, 0.03% CO_2, and 78% nitrogen gas (N_2). When humans breathe gases that are not normally found in air, or in unique concentrations, the body may or may not be able to respond appropriately. Thus, breathing the gas mixture from a pressurized scuba tank presents unique stressors to the respiratory and circulatory systems of the body.

The term *mixed gas* refers to any breathing mixture other than air. *Mixed-gas diving* traditionally refers to diving with mixtures of helium and oxygen (known as *heliox*), air enriched with oxygen (known as *nitrox*), or a blend of oxygen, helium, and nitrogen (known as *trimix*).[21] Nitrogen and helium are not used by the body in respiration; they are **inert gases**. The two main purposes for using special mixed gases in diving are to be able to (a) change the inert gas and (b) control the O_2 level.

Oxygen is always present in diving gases, but it may be a very small fraction of the mixture used for deep dives. It is imperative that the oxygen content of a gas mixture be appropriate for each dive, because O_2 is poisonous at high concentrations. Therefore, oxygen management is a major element of diving safety. Inert gases (e.g., nitrogen, helium, neon, argon, hydrogen) are added to scuba tanks to safely lower the oxygen content.

Seven clinical problems are associated with breathing mixed gases: hypoxia, oxygen toxicity, nitrogen narcosis, decompression sickness, bone necrosis, carbon dioxide toxicity, and breathing gas contamination. These medical conditions are described below.

Hypoxia

Hypoxia refers to inadequate oxygen delivery to the body's tissues; this impairs aerobic metabolism. Overwhelmingly, the greatest hazard pertaining to O_2 in mixed gases is not having enough of it. Although hypoxia is not a concern for most divers, or a major problem in air diving, mixed-gas diving introduces the possibility of getting a mixture without adequate O_2. This can result from breathing the wrong mixture, or from breathing the right mixture at the wrong pressure.[21]

Hypoxia may cause debilitation, unconsciousness, and even death in extreme cases. Ironically, hypoxia can delude divers by making them euphoric and unconcerned about their welfare.

Oxygen Toxicity

It is paradoxical that oxygen is vital to life, but may become a poison to the central nervous system (CNS), lungs, and eyes when present at high levels and for a long duration. Yet, the harmful biochemical effects of **oxygen toxicity** on cell membranes, enzymatic reactions, and metabolism are well known.[22] The increased pressure due to great depth (Boyle's law) is probably the most common way that oxygen toxicity occurs. For example, if a 21% O_2 mixture is breathed at an ocean depth of 91 m (300 ft), external hydrostatic pressure increases the density and decreases the volume of this gas, making it equivalent to breathing 100% O_2 at a depth of only 11 m (36 ft).[23] It is unusual, however, to breathe air at depths below 61 m (200 ft) because of problems with CO_2 toxicity and nitrogen narcosis (see paragraphs below). Obviously, breathing pure oxygen for a prolonged time is a second way to induce oxygen toxicity. Because closed-circuit scuba systems utilize a pure oxygen canister, the potential for oxygen poisoning exists; except in extraordinary circumstances, a closed-circuit scuba apparatus should not be used at a depth of more than 7.6 m (25 ft). Selected symptoms of oxygen toxicity, categorized by the organs affected, are presented in table 4.1.[3, 23, 24] The lungs experience the highest O_2 concentrations of any organ in the body; pulmonary symptoms of oxygen toxicity usually begin with either coughing, a mild irritation under the sternum, or a burning sensation in the trachea and bronchial tubes. If the brain and the CNS are affected,

Table 4.1
Selected Signs and Symptoms of Oxygen Toxicity That Occur During Diving[3,23,27]

Organ/site	Signs and symptoms
Lungs	Chest discomfort during breathing, persistent coughing
CNS*	Apprehension, pale face, sweating, slow heart rate, depression, vertigo, abnormal respiratory function, nausea, vomiting, convulsions, quivering lips, facial twitching, tunnel vision, syncope

* Central nervous system, including the brain.

convulsions are likely, but repeated studies in animals and humans have demonstrated that convulsions are not inherently harmful.[23] The circumstances associated with underwater convulsions *are* dangerous, however. At the height of a seizure, all of the muscles of the body become rigid and the diver's mouthpiece is ejected, leading to death by drowning in most cases. The most common warning signals are muscular twitching, nausea, dizziness, and abnormalities of vision or hearing.[5] Except in cases of severe pulmonary involvement, the treatment for this illness consists of breathing air at sea-level pressure.[3]

Nitrogen Narcosis

You will recall that nitrogen gas (N_2), which constitutes approximately 78% of air, is an inert gas that is not utilized biochemically in respiration. Nevertheless, N_2 in the lungs moves across the alveolar membrane into the blood, and dissolves in the intracellular fluid. You also should recall that the total pressure of each gas in a diver's lungs increases in direct proportion to the depth of the dive. Thus, at a depth of 91 m (300 ft), when lung air equilibrates with intracellular fluid, the tissue N_2 is about three times greater than at the surface.[7] This increased level of N_2 causes many exotic physical and mental symptoms that are similar to alcohol intoxication. Known as **nitrogen narcosis**, this condition generates anesthetic-like euphoria, overconfidence, poor judgment, and a slower reaction time. Many divers have died from nitrogen narcosis because of serious errors in diving techniques and accidents.[3] The greatest hazard may be that it keeps a diver from caring about the task at hand or his or her own safety. Because this condition usually becomes a major problem below 30 m (100 ft), authorities recommend that sport divers not descend below this depth.[3, 7] Interestingly, breathing other inert gases creates a similar euphoric state, known as **inert gas narcosis**. The order of potency of the inert gases, from least to most narcotic, is helium, neon, hydrogen, nitrogen, argon, krypton, and xenon.[25] Because this list is identical to the relative solubility of these gases in lipids, a mechanism for inert gas action has been advanced.[26] This theory proposes that the lipid component of cell membranes is altered by absorption of inert gases into the membrane's molecular structure.[3]

Decompression Sickness

Tissue nitrogen also plays an important role in **decompression sickness** (DCS), known for years as the "bends." The term "bends" originated at the turn of the 20th century among workers who were building the Brooklyn Bridge in New York City. Hobbled by severe pain, these laborers walked by bending forward at the hips because of joint stiffness.[3] DCS typically begins at depths below 30 m (>100 ft; >3 atm), when a diver's tissues are loaded with increased quantities of O_2 and N_2, as described in the previous two paragraphs. Usually the diver returns to the surface too rapidly (swift decompression), causing the gas pressure *within* the tissue to exceed the *external* hydrostatic pressure. This establishes a state of supersaturation and liberates N_2 bubbles within cells, similar to the bubbles that form in a can of carbonated beverage (the air pressure above the soda decreases rapidly when the can is opened).[27] Interestingly, experiments have shown that N_2 bubbles are common in the bloodstream of recreational divers immediately after underwater excursions, and are presumed to be harmless in small volumes because they are filtered by the lung's capillary network and subsequently exhaled.[27] In contrast, excessive N_2 gas bubbles can disturb organ and cell function by (a) blocking arteries, veins, and lymph vessels; (b) causing **compartment syndrome** in a muscle bounded by fascia; and (c) rupturing cell membranes.[28] Signs and symptoms of DCS (table 4.2) occur when decompression is inadequate, within 3 h of completing a dive.[3, 5] When symptoms appear in the CNS, respiratory system, or circulatory system, the required medical treatment is more extensive than if only joint

Table 4.2

Signs and Symptoms of Decompression Sickness ("the Bends") at Various Sites in the Body[3]

Location	Signs and symptoms
Arm and leg joints*	Numbness, dull aching joint pain resembling tendinitis, grating sensation during movements
Entire body	Profound fatigue, generalized heaviness, dizziness
Skin	Mottled rash, itching
Lungs	Burning pain under the sternum, breathlessness, nonproductive cough
Central nervous system	Low back pain, subjective heaviness in the legs, leg numbness, paralysis

* This pain appears in the arms in 70% of all cases.[5]

pain occurs.[5] For example, DiLibero and Pilmanis[29] reported that 24% of scuba divers with DCS suffered some degree of spinal cord injury.

The appearance of symptoms can be avoided if the rate of a diver's ascent (the decompression) is controlled by stopping to rest at various stages of the ascent. If a diver ascends at a slow rate, the risk of getting bubbles in the circulation is lower, because most of the body's excess N_2 will diffuse from the tissues into the blood and be exhaled through the lungs.[27] Years of U.S. Navy research have resulted in the design of decompression schedules that describe safe returns from deep dives; these tables provide rates of ascent, depths, and the duration of rest periods.[5, 7] As an illustration, a dive to 30 m (100 ft) for 50 min requires two decompression stops: one at 6 m (20 ft) for 2 min and another at 3 m (10 ft) for 24 min. This preventive technique has been named **in-water decompression**. Medical treatment at the surface consists of lying in a compression chamber, of which there are several types ranging from a one-person, plastic, portable variety to a larger facility capable of housing several individuals for a longer period of time.[30] The patient usually is compressed to a simulated depth of 18 m (60 ft, 2.8 atm) and breathes 100% O_2. To reduce the risk of oxygen toxicity, the decompression period is interspersed with short intervals of breathing air. Figure 4.9 predicts whether decompression treatment (either in water or in a chamber) will be required or not.

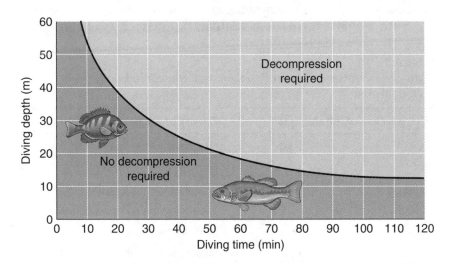

Figure 4.9 The need for decompression treatment can be predicted by using this figure. Any dive falling to the left of the curve requires no decompression, as long as the rate of ascent does not exceed 18.3 m/min (60 ft/min).
Adapted from McArdle, Katch, and Katch 1996.

Bone Necrosis

Bone necrosis, medically known as dysbaric osteonecrosis, is a recognized occupational health hazard among commercial divers.[31] Segments of bone die at a slow rate and may not produce symptoms or appear on X rays for months or years. This disease is most common among divers who have a history of decompression sickness, who frequently dive below 30 m (100 ft), and who ascend without appropriate decompression stops. It is rare in divers who stay within the limits of the U.S. Navy decompression schedules.[5, 31] Bone necrosis occurs most often at the shoulder (humeral head), hip (femoral head), and knee joints (articulation of the femur and tibia).[32]

Carbon Dioxide Toxicity

Chronic elevation of carbon dioxide levels (**CO_2 toxicity**) in the blood most often occurs with a closed-circuit scuba apparatus in which the chemical "scrubbing" efficiency of the CO_2 filter decreases over time. This usually occurs when underwater activity lasts longer than 4-6 h, and is currently recognized by the U.S. Navy as one liability of employing closed-circuit systems.[33] CO_2 toxicity is very rare in open-circuit scuba diving.[3] Although CO_2 is a product of metabolism, and is not a toxin in the traditional sense, elevated levels in human tissues produce the signs and symptoms shown in table 4.3. These effects range from a mild increase in ventilatory rate to loss of conscious-

Table 4.3
Signs and Symptoms of Acute Exposure to Elevated Carbon Dioxide Levels in Normal Men

CO_2 percent of inhaled gas*	Effects of exposure
0-4	No CNS derangement
4-6	Breathlessness, anxiety
6-10	Impaired mental capabilities
10-15	Severely impaired mental function
15-20	Loss of consciousness
>20	Muscular twitching, convulsions

* At sea level. At great depths, the effects of these CO_2 levels will be magnified.
Adapted from Clark and Thom 1997.

ness (10-15% CO_2 in inspired air) and convulsions (>20% CO_2), depending on the level inspired.

Recent research in our laboratory evaluated individual perception of CO_2 in gas mixtures to determine if scuba divers could sense small elevations in the CO_2 content of inspired air, and to provide cues that scuba divers might use to warn themselves of **hypercapnia** (elevated blood CO_2 levels) before it reaches injurious levels.[33] Resting college-aged males could sense that inhaling room air was different from both a 6% and 8% CO_2 gas mixture, and that their responses (respiratory rate, headache, restlessness, faintness, and breathlessness) during the 8% experiment were greater than during the 6% trial. We concluded that military, technical, and labor personnel should avoid tasks that require mental acuity (e.g., air traffic control) for at least 2 h after breathing an 8% CO_2 gas mixture. Further, we suggested that divers use headache as a warning signal of hypercapnia because it occurred more often than any other symptom.

One final physiological effect of CO_2 deserves consideration. A low carbon dioxide level in blood has been implicated in breath-hold diving blackout via hypoxia; this life-threatening syndrome was described above, in the section titled "Predive Hyperventilation: Useful but Risky," page 143.

Breathing Gas Contamination

The final medical condition of this chapter involves **breathing gas contamination**.[23] As scuba gas cylinders are pressurized or filled, the sea-level pressure of each individual gas is multiplied. This means that any contaminant in the air source can become dangerous to the diver. For this reason, compressor motors must be free of oil that could be pumped into tanks; otherwise, the diver may suffer a lung inflammation (e.g., lipoid pneumonitis). Also, the air inlet pipes of compressors must be placed to avoid engine exhaust and carbon monoxide (CO). The hazards of CO inhalation will be described fully in chapter 6 (section titled "Carbon Monoxide" page 201), including its binding affinity for blood hemoglobin, uptake, and elimination. Because CO is colorless, odorless, and tasteless, a diver cannot detect it unless (a) other contaminants exist or (b) symptoms arise (e.g., headache, nausea, dizziness). Unfortunately, these symptoms are similar to those of DCS and air embolism (see above). Fortunately, the best medical treatment for CO poisoning (namely, hyperbaric oxygen administration) is identical to that for DCS and air embolism.[3]

© Norbert Wu/www.norbertwu.com

Divers must keep their air sources clean since even small amounts of contaminents in pressurized breathing gas can be dangerous.

MEDICAL CONSIDERATIONS: CARDIAC PROBLEMS AND SUDDEN DEATH

A number of factors associated with diving increase the likelihood of abnormal electrical impulse transmission and aberrant rhythm in the heart (cardiac arrhythmia). These include the blood pressure increase during breath-hold diving, wet-suit pressure placed on the carotid artery space of the neck, and fatigue.[8] One recent study suggested that heart abnormalities may occur in healthy, young persons.[34] During 24 h monitoring of the electrocardiograms (ECG) of 20 young adult scuba divers, a variety of cardiac arrhythmias were observed. The frequency of arrhythmia while submerged was more than 22 times greater than it was out of the water. The authors suggested that external hydrostatic pressure displaced blood to the center of the body and expanded the heart chambers, altering the cardiac rhythm.[35] But they could not verify that such arrhythmias are the cause of sudden death during diving. Further studies are needed to explore this relationship.[36]

COUNTERACTING BAROTRAUMA

Barotrauma is always a concern among divers, but it can usually be avoided with appropriate precautions.

Hyperbaric Acclimatization

Various sections of this book describe acclimatization to environmental extremes (heat, cold, air pollution), during repeated exposures over a prolonged period of time. Likewise, repeated exposure to the high pressure, low O_2, or high CO_2 (hypercapnia) associated with breath-hold diving produces adaptations in divers that reduce stress and enhance performance. In breath-hold diving, this **hyperbaric acclimatization** manifests itself as a change in lung capacity and enhanced sensitivity to changes in blood constituents.[26] For example, a study of U.S. Navy divers during their first year of duty showed a 4% increase in total lung capacity and a decreased response to elevated CO_2 levels in blood.[37] Another publication reported that respiratory acclimatization to chronic hypercapnia required about 3-5 days, but most of this response occurred in the first 24 h and the adaptations were subtle.[38]

Many authorities propose that acclimatization also occurs when gas mixtures are inhaled.[5] Scuba divers who have had no recent exposure to nitrogen-containing gas at depths below 37 m (122 ft) may be susceptible to nitrogen narcosis. According to one popular theory, a diver gradually develops an increased tolerance[5] to the narcotic effects of N_2. However, there is no objective evidence of this phenomenon, suggesting that it is merely the result of psychological adjustments.[9]

Special Concerns for Female Divers

Participation of women in sport scuba diving increases every year. Current estimates place the percentage of female divers at 33-35% of all participants. Accidents and injuries occur at the same rate in female and male divers, attesting to the effectiveness of credentialing programs. Unfortunately, scientific and medical investigations have paid little attention to gender differences in anatomical, physiological, and psychological factors. Also, the number of studies examining unique female responses to hyperbaric environments are insufficient.

In terms of thermoregulation, women possess a smaller muscle mass than men and generate less metabolic heat during exercise. Some laboratory findings place women at a disadvantage in terms of tolerance to cold-water immersion.[18] However, endurance swimming events offer contradictory evidence; open-water records for distance and time are often held by women. Even if women are more vulnerable to hypothermia, cold stress should not pose a sex-specific hazard for female divers who are properly equipped. Also, the lower $\dot{V}O_2$max and upper body strength of women should have little influence on their risk of injury or performance. Clearly, well-trained, active people can dive safely, regardless of gender.[39]

Women and the Diving Illnesses

You will recall that the solubility of N_2 in fat is relatively great. This prompted the theory that women may be more susceptible to some diving illnesses than men, because women typically have more subcutaneous body fat. But most field studies do not report a female versus male difference in the incidence of nitrogen narcosis, decompression sickness (DCS), or oxygen toxicity.[39]

However, female susceptibility to DCS may increase during certain phases of the menstrual cycle. Three compression chamber studies have reported that the number of DCS symptoms during menstruation were greater than during other times of the month.[40, 41, 42] Further, premenstrual syndrome (PMS), a poorly defined condition, may alter mood and personality in some women during the days prior to onset of menses. When *severe* PMS affects a female diver, it is unwise for her to dive because no one at risk for depressive or antisocial tendencies should dive.[39] Oral and injectable contraceptives do not appear to increase the risk of hyperbaric injury for women who dive.[39]

Diving During Pregnancy

Scientific evidence indicates that a hyperbaric environment may adversely affect fetal development in pregnant women who dive. The concerns for fetal health are as follows:

• Two published surveys provide uncertain results regarding the effects of diving on the outcome of pregnancy. In her survey of 72 women, Bangasser found no increase in the rates of birth defects or stillbirths when pregnant divers were compared to pregnant nondivers.[43] In contrast, Bolton surveyed 109 women who dived before and during gestation; 69 stopped when their pregnancy was recognized and 40 continued diving during fetal development.[44] The results suggested higher rates of low birth weight, respiratory difficulties, and birth defects (e.g., defects of multiple vertebrae, absence of a hand, and heart defects involving the ventricle, aorta, and a heart valve) in the group who continued to dive as the fetus grew.

• Decompression during ascent may place a fetus at increased risk.[39] This is significant because the fetal cardiovascular system lacks an effective filter, and bubbles are likely to be directed to the brain (causing stroke) and coronary arteries (causing cardiac arrest).

• Although speculative, one authority[39] expressed concern that the physiologic changes caused by pregnancy might compound the risks of diving in three ways. First, combining the increased water and fetal weight, exercise demand of diving, cold-water immersion, and stress of anxiety, may increase the possibility of uterine vasoconstriction. Such episodic demands are more likely to compromise uterine blood flow than the sustained requirements of aerobic activity. Second, during the early months of pregnancy, approximately two-thirds of pregnant women experience some degree of gastrointestinal dysfunction (e.g., nausea, vomiting, increased gastric acid). If motion sickness on the dive boat adds to morning sickness, the pregnant diver will be at high risk of vomiting into her breathing regulator. This is an accident that few sport divers are prepared to handle safely. Third, maternal body water is redistributed; interstitial fluid (between blood vessels and cells) and edema/swelling increase. This fluid is a potential reservoir for nitrogen, as is maternal body fat.

When one weighs the elective nature of sport diving against the possibility of a lifelong disability created in utero by pressurized gas, the rational decision seems clear. Pregnant women should not dive.[39] This has been advanced as an official position of the British Sub Aqua Club, the Undersea Medical Society, and the major diving organizations in America.[8, 45]

UNDERWATER EXERCISE PERFORMANCE

This chapter has described very few matters concerning human performance. The nature of diving is such that exercise intensities are rarely maximal due to the risks and respiratory limitations inherent in both breath-hold and scuba diving, and because of the required waiting periods during ascent from great depths. Although brief, strenuous bursts of activity are occasionally required, the major limitations on underwater performance are often viewed as psychological or perceptual, not physiological.[15] Because the majority of this chapter has dealt with illnesses and injuries that are associated with diving, this sport may appear to be dangerous or risky. Clearly, the technical aspects of mixed gases and underwater pressures may be daunting to some individuals, but, when done safely, sport diving can provide great enjoyment and many rewards. If you follow the safety guidance below, your risk of diving illness or injury will be minimized.

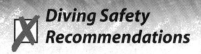

Diving Safety Recommendations

Recreational diving is usually done at a site that is hours or days away from the nearest recompression chamber, so DCS usually requires a major evacuation effort. Further, several factors confound any set of decompression tables, including workload during the dive, water temperature, individual variability, and postdive exercise or altitude exposure.[3,8] Considering the potentially devastating consequences of DCS, recreational and sport divers should always adhere to the following safe diving recommendations, written by Dr. Kenneth W. Kizer:[3]

- Dive using depths and durations that are within the limits of "no decompression," as shown in figure 4.9.

- Ascent after completing rest stops (as prescribed in decompression schedules) should be accomplished at a rate of 0.3 m/sec (1 ft/sec) or slower.

- After any dive to 18 m (60 ft) or more, include a safety stop of 3 to 5 min at a depth of 3-4.5 m (10-15 ft).

- Carefully plan each repetitive dive so it will be shallower than previous dives.

- Do not allow your body to become dehydrated. If you remain well hydrated, capillary perfusion will be enhanced and inert gas exchange will occur optimally. This will decrease the risk of DCS. Follow the guidelines in chapter 2 for avoiding dehydration. Also, use the urine color chart on the rear cover of this book to check your daily hydration status.

- To decrease the risk of DCS, do not engage in strenuous exercise, such as windsurfing or jogging, for at least 6 h after a dive.

- Do not fly in an aircraft, even in the pressurized cabin of commercial airlines, for at least 12 h after diving. If in-water decompression stops were required, wait at least 24 h before flying.

- Diving in mountain lakes requires major adjustments in decompression tables to account for the decreased atmospheric pressure at the surface of the lake. This will take on greater meaning in chapter 5, which focuses on high to extreme altitude environments.

REFERENCES

1. Walker, D. 1836. *British manly exercises. Containing rowing and sailing, riding and diving, etc.* Philadelphia: T. Wardle, 87-88.

2. Vorosmarti, J. 1996. Hyperbaria/diving: Introduction. In *Handbook of physiology. Section 4: Environmental physiology.* Vol. II, edited by M.J. Fregly & C.M. Blatteis, 975-978. New York: Oxford University Press.

3. Kizer, K.W. 1995. Scuba diving and dysbarism. In *Wilderness medicine. Management of wilderness and environmental emergencies,* 1176-1208. St. Louis: Mosby.

4. Nukada, M. 1965. Historical development of the Ama's diving activities. In *Physiology of breath-hold diving and the Ama of Japan,* edited by H. Rahn & T. Yokoyama. Washington, DC: National Academy of Sciences, National Research Council.

5. U.S. Navy. 1985. *U.S. Navy diving manual. Air diving.* Vol. 1, 1-18. Washington, DC: U.S. Government Printing Office.

6. Craig, A.B. 1967. Summary of 58 cases of loss of consciousness during underwater swimming and diving. *Medicine & Science in Sports:* 171-175.

7. McArdle, W.D.; Katch, F.I.; & Katch, V.L. 1996. *Exercise physiology: Energy, nutrition, and human performance.* Baltimore: Williams & Wilkins, 526-537.

8. Strauss, R.H. 1984. Medical aspects of scuba and breath-hold diving. In *Sports Medicine,* edited by R.H. Strauss, 361-377. Philadelphia: Saunders.

9. Lundgren, C.E.G.; Harabin, A.; Bennett, P.B.; Van Liew, H.D.; & Thalmann, E.D. 1996. Gas physiology in diving. In *Handbook of physiology.* Section 4: *Environmental physiology.* Vol. II, edited by M.J. Fregly & C.M. Blatteis, 999-1019. New York: Oxford University Press.

10. Beebe, W. 1934. *Half mile down.* New York: Duel, Sloan, & Pierce.

11. Lin, Y.C., & Hong, S.K. 1996. Hyperbaria: Breath-hold diving. In *Handbook of physiology.* Section 4: *Environmental physiology.* Vol. II, edited by M.J. Fregly & C.M. Blatteis, 979-995. New York: Oxford University Press.

12. Teruoka, G. 1932. Die Ama und ihre arbeit. *Arbeitsphysiologie 5:* 239-251.

13. Hong, S.K.; Henderson, J.; Olszowka, A.; Hurford, W.E.; Falke, K.J.; Qvist, J.;

Radermacher, P.; Shiraki, K; Mohri, M.; Takeuchi, H.; Zapol, W.J.; Ahn, D.W.; Choi, J.K.; & Park, Y.S. 1991. Daily diving patterns of Korean and Japanese breath-hold divers. *Undersea Biomedical Research 18*: 433-443.

14. Kang, D.H.; Kim, P.K.; Kang, B.S.; Song, S.H.; & Hong, S.K. 1965. Energy metabolism and body temperature in the ama. *Journal of Applied Physiology 20*: 46-50.

15. Egstrom, G.H., & Bachrach, A.J. 1997. Human performance underwater. In *Diving medicine*, 3rd ed., edited by A.A. Bove, 77-88. Philadelphia: Saunders.

16. Lin, Y.C. 1988. Applied physiology of diving. *Sports Medicine 5*: 41-56.

17. Schneider, E.C. 1930. Observations on holding the breath. *American Journal of Physiology 94*: 464-470.

18. Hong, S.K. 1997. Breath-hold diving. In *Diving medicine*, 3rd ed., edited by A.A. Bove, 65-74. Philadelphia: Saunders.

19. Craig, A.B. 1961. Causes of loss of consciousness during underwater swimming. *Journal of Applied Physiology 16*: 583-586.

20. Green, S.M.; Rothrock, S.G.; Green, E.A. 1993. Tympanometric evaluation of middle ear barotrauma during recreational scuba diving. *International Journal of Sports Medicine 14*: 411-415.

21. Hamilton, R.W. 1997. Mixed-gas diving. In *Diving medicine*, 3rd ed., edited by A.A. Bove, 38-64. Philadelphia: Saunders.

22. Haugaard, N. 1968. Cellular mechanisms of oxygen toxicity. *Physiological Reviews 48*: 311-373.

23. Clark, J.M., & Thom, S.R. 1997. Toxicity of oxygen, carbon dioxide, and carbon monoxide. In *Diving medicine*, 3rd ed., edited by A.A. Bove, 131-145. Philadelphia: Saunders.

24. Donald, K.W. 1947. Oxygen poisoning in man. *British Medical Journal 1*: 712-717.

25. Bennett, P.B. 1975. Inert gas narcosis. In *The physiology and medicine of diving and compressed air work,* edited by P.B. Bennett & D.H. Elliott, 207-230. London: Bailliere Tindall.

26. Muza, S.R. 1988. Hyperbaric physiology and human performance. In *Human performance physiology and environmental medicine at terrestrial extremes,* edited by K.B. Pandolf, M.N. Sawka, & R.R. Gonzalez, 565-589. Indianapolis: Benchmark Press.

27. Moon, R.E.; Vann, R.D.; & Bennett, P.B. 1995. The physiology of decompression illness. *Scientific American 273*: 70-77.

28. Melamed, Y.; Shupak, A.; & Bitterman, H. 1992. Medical problems associated with underwater diving. *New England Journal of Medicine 326*: 30-35.

29. DiLibero, R.J., & Pilmanis, A. 1983. Spinal cord injury resulting from scuba diving. *American Journal of Sports Medicine 11*: 29-33.

30. Murrison, A.W., & Francis, T.J.R. 1991. An introduction to decompression illness. *British Journal of Hospital Medicine 46*: 107-110.

31. Elliott, D.H. 1997. Medical evaluation for commercial diving. In *Diving medicine*, 3rd ed., edited by A.A. Bove, 361-371. Philadelphia: Saunders.

32. Elliott, D.H., & Harrison, J.A.B. 1971. Aseptic bone necrosis in Royal Navy divers. In *Underwater physiology*, edited by C.J. Lambertsen, 251-262. New York: Academic Press.

33. Maresh, C.M.; Armstrong, L.E.; Kavouras, S.A.; Allen, G.J.; Casa, D.J.; Whittlesey, M.; & La Gasse, K.E. 1997. Physiological and psychological effects associated with high carbon dioxide levels in healthy men. *Aviation, Space, and Environmental Medicine 68*: 41-45.

34. McDonough, J.R.; Barutt, J.P.; & Saffron, J.C. 1987. Cardiac arrhythmias as a precursor to drowning accidents. In *The Physiology of breath-hold diving*, Publication no. 72, 212-219. Bethesda, MD: Undersea and Hyperbaric Medical Society.

35. McDonough, J.R.; Barutt, J.; & Saffron, J.C. 1992. Letter to the editor. *New England Journal of Medicine 326*: 1998.

36. Melamed, Y.; Shupak, A.; Bitterman, H.; & Weiler-Ravell, D. 1992. Reply to a letter to the editor. *New England Journal of Medicine 326*: 1998.

37. Schaefer, K.E. 1965. Adaptation to breath-hold diving. In *Physiology of breath-hold diving and the Ama of Japan*, edited by H. Rahn & T. Yokoyama, 237-252. Washington, DC: National Academy of Sciences, National Research Council.

38. Clark, J.M.; Sinclair, R.D.; & Welch, B.E. 1971. Rate of acclimatization to chronic hypercapnia in man. In *Underwater physiology*, edited by C.J. Lambertsen, 399-407. New York: Academic Press.

39. Taylor, M.B. 1997. Women in diving. In *Diving medicine*, 3rd ed., edited by A.A. Bove, 89-107. Philadelphia: Saunders.

40. Dunford, R.G., & Hampson, N.B. 1992. Gender-related risk of decompression sickness in hyperbaric chamber attendants: A case control study. Abstract. *Undersea Biomedical Research 19*: 37.

41. Rudge, F.W. 1990. Relationship of menstrual history to altitude chamber decompression sickness. *Aviation, Space, and Environmental Medicine 60*: 657-659.

42. Dixon, G.; Krutz, R.; & Fischer, J. 1988. Decompression sickness and bubble formation in females exposed to a simulated 7.8 psi suit environment. *Aviation, Space, and Environmental Medicine 59*: 1146-1149.

43. Bangasser, S.A. 1978. Medical profile of the woman scuba diver. In *Proceedings of the 10th International Conference on Underwater Education*, 31-40. Colton, CA: National Association of Underwater Instructors.

44. Bolton, M.E. 1980. Scuba diving and fetal well-being: A survey of 208 women. *Undersea Biomedical Research 7*: 183-189.

45. Cresswell, J.E., & St. Leger-Dowse, M. 1991. Women and scuba diving. *British Medical Journal 302*: 1590-1591.

chapter 5
Altitude: Earth's Hypobaric Environment

I entered the world of mountaineering in 1988, when I climbed Mount Rainier via the Camp Muir route. There were seven of us, five novice mountaineers and two accomplished guides. Our first day involved a leisurely hike from Paradise Lodge at 1650 m (5413 ft) to Camp Muir at 3050 m (10,007 ft), via the snowfields. After dinner, we slept in our tents until 11:00 P.M., put on our climbing gear, roped up, and went for the summit attempt. The moon was almost full and the alpine glow was spectacular as we ascended Disappointment Cleaver. At about 3660 m (12,008 ft), I began to experience the effects of chronic hypoxia: fatigue, irritability, and a headache in the frontal lobe. As we approached the false summit at 4270 m (14,009 ft), my headache spread to the top of my head and felt like it was securely fastened inside a vice grip. I was very nauseous and wanted to vomit, but did not. My body was extremely dehydrated and my urine was a dark yellow color. After shaking hands and

taking pictures, we started down. The descent was fast. My
headache and nausea went away and the fatigue was less
noticeable.

—Donald Winant, 1998

The previous chapter described earth's only hyperbaric environment. Here we consider the opposite challenge to homeostasis: the reduced pressure (hypobaria) and oxygen content (hypoxia) of high terrestrial environments, known in everyday terms as "thin air." When the body acclimatizes to high altitude, primarily the respiratory and circulatory systems respond to hypobaria and hypoxia. Several unique illnesses may result from high-altitude exposure; their incidence increases with the speed of the ascent, altitude reached, and the length of stay. Safety recommendations focus on nutritional, fluid, and pharmacologic interventions that optimize health and exercise performance.

THE NATURE OF AIR AT HIGH ALTITUDE

The opening description of acute mountain sickness was recorded during a conversation with Donald Winant, who is a determined triathlete as well as a former collegiate swimmer and Air Force officer. He is one of the few people on earth who has had the opportunity to experience hypoxia (low O_2 levels in blood or tissues) in three different ways: flying high-performance jet aircraft (acceleration-induced hypoxia), resting in an altitude chamber (acute hypobaric hypoxia), and climbing a mountain (chronic altitude-induced hypoxia).

In contrast to chapter 4, which focused on *hyperbaria* in underwater environments, chapter 5 deals with the physiologic responses and medical problems associated with **hypobaria** (low pressure) at high terrestrial altitudes. The atmospheric pressure at these heights is considerably less than at sea level (which is known as "1 atmosphere," or "1 atm"). Figure 5.1 illustrates this fact. A young woman is holding a balloon at various locations above and below sea level; the temperature is identical at all locations. At sea level, the volume of the gas inside the balloon is affected by the atmospheric pressure of that location, 1 atm. As she descends into a mine, the volume of the balloon decreases due to increasing air pressure (Boyle's law; see chapter 4). Conversely, as she ascends a mountain, the balloon expands

Figure 5.1 The volume of air inside a balloon changes at different altitudes, above and below sea level. Although the percentage of oxygen is constant, the density of the gas changes due to differences in atmospheric pressure.[1]

as the atmospheric pressure decreases.[1] This is due to the fact that the column of air compressing the balloon *on the mountain* weighs less than that *inside the deep mine*.

Above sea level, the atmospheric pressure diminishes by about 50% for an increase of approximately 5500 m (18,045 ft). Thus, air has a lower density at higher altitudes (i.e., fewer particles per cubic inch) because the gas has expanded. Even though the *percentage* of oxygen and other gases in air does not change as altitude increases (up to an altitude of 100,000 m, 328,083 ft), the "thinner air" presents less oxygen to the lungs, alters physiologic responses, and produces unique illnesses because of the decrease in gas density.

Three definitions will help you understand the physiologic responses and medical problems associated with hypobaric-hypoxic environments (i.e., those with low air pressure and low O_2).[2] In the paragraphs below, the term *high altitude* refers to sites that are 1500-3500 m (4921-11,483 ft) above sea level. *Very high altitude* refers to locations that are 3500-5500 m (11,483-18,045 ft) above sea level; more than 10 million people live at this elevation range.[3] The term *extreme altitude* refers to any locale above 5500 m (18,045 ft); few humans reside above this elevation. It is likely that about 5820 m (19,094 ft) is the upper limit for human habitation,[2] although the caretakers of the Aucanquilcha mine in northern Chile live at 5985 m (19,636 ft).[4] When no specific elevation is implied (or when a principle applies to high, very high, and extreme altitudes), the general term *altitude* is utilized.

YOUR BODY'S RESPONSES TO HIGH TERRESTRIAL ENVIRONMENTS

The distinctive result of exposure to high to extreme altitude is **hypoxia**. In metabolic terms, hypoxia is a state in which the rate of oxygen utilization by cells is inadequate to supply all of the body's energy requirements. This energy deficit may be compensated, within limits, by energy production from **anaerobic metabolism** (i.e., anaerobic glycolysis).[5] But this biochemical pathway has three inherent disadvantages: it yields far less **ATP** per gram of fuel than **aerobic metabolism**; it depends largely on the availability of carbohydrate; and it produces lactic acid that may disturb the acid-base balance of cells.

The body judiciously defends O_2 supply to the brain and other organs by initiating numerous responses to acute hypoxia, which may last a few minutes, hours, or days.[1] These short-term responses are described in the following paragraphs. Although the boundary between short-term *responses* and long-term *adaptations* is not always clear, long-term adaptations will be described later in this chapter in the sections titled "Hypobaric-Hypoxic Environment and Physical Performance" (see page 173), and "Altitude Acclimatization" (see page 178).The body systems shown in figure 5.2 (page 170) respond to hypoxia and reduced air pressure at altitude.

Respiratory System

Pulmonary ventilation increases immediately with exposure to altitude because chemoreceptors in the carotid artery and aorta sense a lower oxygen level in blood. Climbers who scale Mount Everest (over 8840 m or 29,003 ft) and the world's highest peaks are able to resist hypoxia by extreme hyperventilation, which adds more O_2 to the blood.[4] This response also causes CO_2 to be exhaled and the blood pH to rise (i.e., respiratory alkalosis). The greatest ventilatory rates ever recorded were seen in eight climbers on Mount Everest, at an altitude of 6340 m (20,800 ft). Their average respiratory frequency was 62 breaths per minute and their total ventilation was 207.2 L (218.8 qt)/min! This extremely high ventilatory rate is facilitated by the low external air pressure, which reduces the amount of work required during breathing. However, the work of the respiratory muscles during hyperventilation accounts for about 10% of the total resting oxygen uptake at extreme altitude. This is a metabolic burden not experienced at sea level.[4]

Cardiovascular System

Cardiac output increases substantially on arrival at extreme altitude,[4] due to an increase in heart rate.[4, 5] In response to hypoxia, stroke volume decreases for a given work rate,[6] but little or no change occurs in the systolic and diastolic blood pressures. At the same time, the venous return of blood to the heart and cardiac contractility are maintained.[6] Cardiac arrhythmias are uncommon at extreme altitude, but irregular cardiac rhythms and marked cyclic variation of heart rate are very common during sleep if periodic breathing develops (see below).[4] At altitude, release of O_2 in tissue capillaries is enhanced by the unique properties of **hemoglobin**, the molecule that carries O_2 on red blood cells.[5]

Interestingly, **polycythemia** (increased rate of red blood cell production) is often considered to be one of the classic, rapid responses to hypoxia because it supposedly increases the oxygen-carrying capacity of blood. But the increase in hemoglobin that occurs during the first two days at altitude actually is due to a loss of plasma volume.[2] This response also may raise the viscosity of blood, causing uneven blood flow and "sludging" in muscle capillaries.[3] Recent research supports this concept; diluting the blood of polycythemic mountaineers did not impair their exercise performance and induced a small improvement in psychomotor tasks.[7]

The Body's Responses to Stress

Nervous System

Maintains homeostasis:

- hypothalamus receives afferent input
- sends efferent messages that maintain/alter blood gas and acid-base balance
- vision, memory, and sleep deteriorate at high altitude

Cardiovascular System

- oxygen-carrying capacity of the blood increases
- cardiac muscle mass increases
- chemoreceptor (O_2, CO_2) sensitivity increases

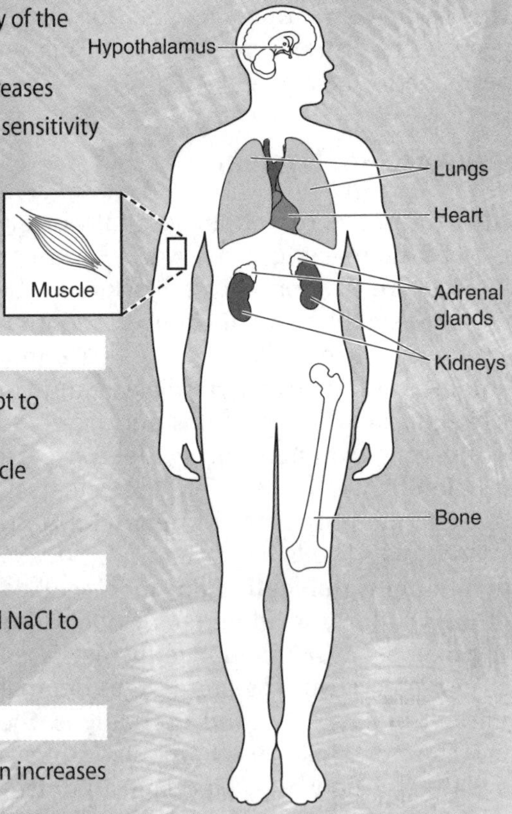

Muscular System

- respiratory muscles adapt to hypobaria
- myoglobin levels in muscle increase

Urinary System

- kidneys retain water and NaCl to offset dehydration

Skeletal System

- red blood cell production increases in bone marrow

Endocrine System

- fluid-electrolyte hormones retain water and NaCl via the kidneys to offset dehydration
- adrenal hormones alter the type of fuels used in muscle and the ratio of aerobic-to-anaerobic metabolism

Respiratory System

$\dot{V}O_2$max decreases

Hypothalamus

Muscle

Lungs

Heart

Adrenal glands

Kidneys

Bone

Figure 5.2 The body systems that adapt to high-altitude environments.

This natural polycythemic adaptation, stimulated by the production of the hormone **erythropoietin (EPO)** in the kidneys,[8] has provided a model for **"blood doping"** (also known as "blood boosting," "blood packing," and "autologous erythrocyte reinfusion"), used illegally by distance runners since at least the Montreal Summer Olympic Games in 1976. Assisted by physicians, they have utilized one of two techniques. First, once the increased red blood cell volume of altitude has been achieved, two units of blood (approximately1 L [1 qt]) are removed; the average adult has 4-5 L of blood in the entire circulatory system. After spinning this blood in a centrifuge, the red blood cells are frozen and reinfused into the circulation of the same athlete 6 weeks later. The goal is to increase the O_2-carrying capacity of blood and to increase the amount of O_2 delivered to exercising muscles. This medical technique was found to improve $\dot{V}O_2$max and exercise performance 24 h after reinfusion, and apparently longer.[8]

The supplemental use of EPO constitutes the second "blood doping" technique.[9] With modern molecular biology techniques (such as recombinant DNA cloning), EPO may now be produced synthetically. Injected synthetic EPO can take up to 3 weeks to fully stimulate red blood cell production in bone marrow. Excessive use of EPO may lead to a dangerously high hematocrit (red blood cell concentration in the plasma). Whereas normal hematocrit ranges from 42 to 46%, EPO-induced hematocrit may be as high as 55%. Because endurance athletes routinely experience dehydration, their plasma volume losses are large and their risk increases for viscous blood, poor circulation (leading to reduced O_2 delivery to muscles), and heart attack.[9] In fact, among elite cyclists, several deaths have been attributed to illegal use of synthetic EPO since it became available in the late 1980s. It also is used illegally by many European distance runners, despite being banned by the International Olympic Committee and the U.S. National Collegiate Athletic Association. Unfortunately, no laboratory technique is available to detect the artificial EPO molecule or to distinguish it from the body's own erythropoietin.

Central Nervous System

Because the CNS is very sensitive to hypoxia, it is not surprising that many changes in neuropsychological function have been observed at extreme altitude. These include alterations in vision, hearing, motor skills, memory, and mood.[4, 5] Visual hallucinations, tunnel vision, and persistent psychiatric disturbances also have been reported.[4, 10] Generally, most authorities agree that accurate work can be performed at extreme altitude, but they require more time and more effort in concentration.

Metabolism

Weight loss is common at extreme altitude, resulting from the factors depicted in figure 5.3. This was demonstrated clearly during the Himalayan Scientific and Mountaineering Expedition of 1960-1961, led by Sir Edmund Hillary and noted scientists. Their research team established a laboratory at 5834 m (19,140 ft) and recorded observations for several months. Most subjects lost between 0.5 and 1.5 kg (1.1 and 3.3 lb) per week.[11] The causes of this weight loss are not certain, and may be different for each climber. However, other features of life at extreme altitude[4] are relevant to this matter: (a) hypoxia and the symptoms of acute mountain sickness reduce appetite; (b) eating and drinking are overshadowed by the need to concentrate on physical tasks and climbing; and (c) protein metabolism may be curtailed.

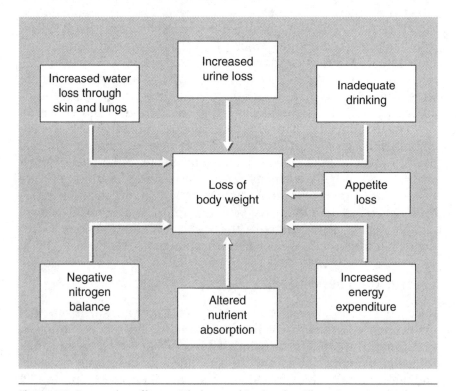

Figure 5.3 Factors that affect weight loss at altitude, when illness is not involved.
Adapted from Cymerman 1996.

Sleep

At high altitude, sleeping is impaired in virtually all humans. Climbers often complain that they have trouble falling asleep, are frequently disturbed in the middle of the night, and do not feel refreshed when they wake in the morning.[2, 4] Increased hypoxic stress, at higher altitude, increases the amount of disturbed sleep.[12] Cheyne-Stokes (periodic) breathing is another characteristic of sleep at altitude. Although some authorities state that periodic breathing is the cause of night disturbance,[4] observations of climbers have shown that this respiratory abnormality may occur with or without this sleep disturbance.[14] Cheyne-Stokes breathing (named after the two scientists who discovered this phenomenon) occurs when hypoxia increases nocturnal breathing rate, CO_2 expiration, and respiratory alkalosis.[2] The increase in pH stimulates the brain's respiratory control center to reduce breathing rate; next, the arterial O_2 content decreases and the blood CO_2 level increases, causing a rise of ventilatory rate and a recurring cycle of disrupted slow and fast breathing.[4] The most effective prescription drug for treating this phenomenon is acetazolamide. It reduces periodic breathing, increases oxygenation, and is a safe sleeping aid.[14]

HYPOBARIC–HYPOXIC ENVIRONMENT AND PHYSICAL PERFORMANCE

Despite a number of investigations and theories, it is not clear what limits exercise performance at high altitudes. Dr. Charles Houston, one of the world's foremost altitude physiologists, explains our current understanding in this way.[15] It seems likely that muscular fatigue limits performance up to about 3000-4600 m (9842-15,092 ft). Between 4600 and 6100 m (15,092 and 20,013 ft), the causes of decreased performance appear to be unique for each individual. Above an elevation of 6100 m, an inability to breathe or to diffuse enough air (i.e., from inside the lung to the blood) are the primary constraints.

Maximal Exercise

Climbers at altitude require a maximal physical effort only on rare occasions. Even then, they can perform only briefly and must take time to recover. Because maximal exercise at extreme altitude requires very high exercise ventilation, this becomes one of the most obvious features of any trek. Near the summit of Mount Everest, for example, every step forward and upward requires 7 to 10 complete respirations. One climber[16] vividly described this experience by stating, "After every few steps, we huddle over our ice axes, mouths agape, struggling for sufficient breath to keep our muscles going. I have the feeling I am about to burst apart. As we get higher, it becomes necessary to lie down to recover our breath."

Above 1500 m (4921 ft), the decline in maximal aerobic power ($\dot{V}O_2$max) due to altitude exposure equals approximately 3% per 300 m; this effect is absent below 1500 m. However, this response varies between individuals. Figure 5.4 shows the range of $\dot{V}O_2$max reduction (from <5 to >25 ml $\cdot$ kg^{-1} $\cdot$ min^{-1}) experienced by 54 sea-level residents who were studied at an elevation of 4325 m (14,190 ft). This variability is likely due to the nature of $\dot{V}O_2$max, which is not determined by a single parameter but involves many physiologic components in a complex O_2 transport system, from the alveoli in lungs to mitochondria in muscle cells. Other important physiologic components include pulmonary ventilation rate, acid-base status, cardiac output, and hemoglobin concentration.[4]

At extreme altitude, the maximal work rate **(power)** and $\dot{V}O_2$max fall rapidly as altitude increases. This translates to poorer perfor-

Mark Newman/Tom Stack & Associates

At high altitudes, even moderate physical exertion requires very high exercise ventilation.

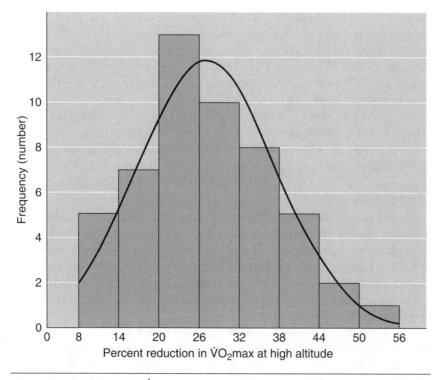

Figure 5.4 The decrease in $\dot{V}O_2$max of 54 sea-level residents, during a visit to 4325 m (14,190 ft). Note that there is a great variability between individuals in the loss of maximal aerobic power (range: <5 to >25 ml · kg^{-1} · min^{-1}).
Reprinted from Young 1988.

mances in competition. The times of track and field and swimming events increase by 3% in events lasting 4 min, and by 8% in events with a duration of 1 h.[17]

Although anaerobic glycolysis compensates for the reduction in aerobic capacity at extreme altitude (see previous section), this response does not occur during maximal exercise because anaerobic metabolism also is greatly restricted.[4] Human muscle biopsies have shown that intramuscular lactate levels are lower during maximal exercise in a hypobaric-hypoxic environment versus exhaustive exercise at sea level.[18]

The Benefits and Disadvantages of Training at Altitude

Much research and debate preceded the 1968 Summer Olympic Games in Mexico City (elevation 2286 m, or 7500 ft). In fact, the U.S. Olympic

Track and Field team trained in Alamosa, Colorado, at an altitude similar to that of the Olympic venue.[3] Those endeavors focused on practical questions such as, "Do those athletes who train at altitude perform better at altitude than those who do not (or those who live at sea level)?" and, "Where should competitors train to prepare for events held at high altitude?" Unfortunately, only a few firm conclusions could be drawn. To understand this research, a few definitions are necessary. Those individuals who stay at altitude for several days or weeks are named *visitors*; those who live there for months or years are *residents*. Persons who were born at altitude are known as *natives*, and those who live permanently at sea level or low altitudes are named *lowlanders*.[15]

Studies suggest that anaerobic (i.e., strength and power) sports are not necessarily affected by high-altitude exposure.[3] Thus, track and field *sprinters* gain nothing, or may lose speed, by training at altitude. In *field events* of shorter duration that involve less exertion, the individual effects vary widely. *Distance runners* usually improve their times after many weeks of training at altitude, but do not match high-altitude natives.[15]

Optimal training for increased performance at high altitude depends on the athlete's altitude of residence and the athletic event. For aerobic events lasting more than 3 or 4 min at altitudes above 2000 m (6562 ft), acclimatization for 10 to 20 days is necessary.[19] $\dot{V}O_2$max was shown to improve slightly (+2.6%) after 20 days of training at altitude.[20] For events conducted above 4000 m (13,123 ft), acclimatization at an intermediate altitude is recommended to allow time for adaptations to develop fully and without illness. Highly anaerobic events conducted at high altitude require only arrival at the time of the event, although acute mountain sickness (see page 183) may become a problem.[2]

Strenuous endurance training at sea level increases $\dot{V}O_2$max. However, at altitude, $\dot{V}O_2$max increases slightly or not at all due to endurance training. But the response of endurance time to exhaustion at altitude is different; it improves greatly. When healthy young males cycled at 75% $\dot{V}O_2$max, their endurance capacity increased 61% after a 12-day stay at 4325 m (14,190 ft).[21] In another investigation, treadmill running time to exhaustion (at 85% $\dot{V}O_2$max) increased 41% after 16 days at the same elevation.[22] Sedentary altitude acclimatization also increases submaximal endurance exercise performance at altitude, but usually has little or no effect on $\dot{V}O_2$max, as shown in table 5.1.[23]

One disadvantage of training at altitude is that exercise intensity must be reduced.[8] For example, distance runners who reside at 4000 m

Table 5.1
Sedentary Altitude Acclimatization Has Little or No Effect on Maximal Aerobic Power ($\dot{V}O_2$ max)

Altitude (m)	Duration of stay (weeks)	Effect on $\dot{V}O_2$ max
3118	3	No change
3822	5	4% increase
4325	2	No change
4325	3	10% increase
4325	3	No change
4325	3	No change or decrease

Note: Each line of data represents a different group of subjects.

Reprinted from Young 1988.

(13,123 ft) may have to reduce their training intensity to 40% of that at sea level (i.e., a much slower pace) during the first 3 weeks to complete their workouts. Over the next 8 weeks at altitude, the duration and intensity may be increased gradually, to 75% of sea-level training.[24] This scenario certainly would place elite endurance athletes who trained at altitude at a disadvantage when competing against

© Jurgen Ankenbrand/J.A. Photographics

Distance runners who train at high altitudes may improve their times in high-altitude events.

athletes who had been training at sea level at a faster pace. However, high-intensity "interval" training (fast runs repeated over a short distance with brief rest periods) is not detrimentally affected by hypoxia, and may be enhanced.[26] Therefore, during early altitude acclimatization, athletes may find it beneficial to emphasize this type of training to maintain muscle power and to offset reductions in endurance training.[26, 27]

Does Altitude Training Enhance Competition?

Some athletes train at high altitude to enhance their performance at sea level. This obviously requires acclimatization. Selection of such a training site should consider the altitude that (a) allows optimal physiologic adaptations and (b) minimizes the decrease of $\dot{V}O_2$max and performance due to hypoxia (at altitudes greater than 1500 m or 4921 ft). Hence, training above 2400 m (7874 ft) apparently affords no increase in subsequent sea-level performance, perhaps for the reasons delineated in the previous paragraph.[2]

Recent research suggests that training benefits occur when endurance athletes are exposed intermittently to altitude.[28] However, disagreement exists on this point because other authorities believe (based on several research studies)[3,8,26,31] that training at altitude provides no or few benefits for sea-level competition beyond those derived by training at sea level. In contrast, there is little disagreement about the superiority of high-altitude natives and residents. The Mexico City Summer Olympics of 1968 provided an excellent example. The first five runners in the 10,000 m (6.21 mi) run were either natives or residents of altitude.[3] Further evidence stems from the recent domination of long-distance events by Kenyan distance runners.[15]

It is also important that you realize that training at sea level benefits sea-level exercise performance when combined with acclimatization (living) or sleep at moderate altitude. Two different research teams have reported that these strategies result in improved performance and physiological responses, such as increased $\dot{V}O_2$max and expanded red blood cell mass (enhanced oxygen-carrying capacity in the blood).[28, 30]

ALTITUDE ACCLIMATIZATION

Rapid ascent to the summit of Mount Everest (>8840 m, 29,000 ft) could cause loss of consciousness and death. Yet climbers ascending this peak over a period of weeks, without supplemental O_2, have experienced only minor medical problems. The process by which humans gradually adjust to hypoxia, enhance performance, and increase their chances of survival is known as **altitude acclimatization**. Successful acclimatization also improves sleep and protects against altitude illnesses.[2]

Englishman Doug Scott—who has successfully conquered Mount Everest, Mount McKinley, and several of the world's tallest peaks—commented on the importance of acclimatization. "No, I don't acclimatize quickly, but once I have acclimatized, I seem to get along well. I have to allow myself a good three weeks before I feel like climbing reasonably, before I feel anything like climbing. . . . I don't know anybody who won't benefit from three weeks of acclimatizing."[32] His estimate of the time required to acclimatize has been supported by several experts.[5, 8, 26, 33]

Numerous physiologic adaptations occur in humans during chronic exposure to a hypobaric-hypoxic environment such as Mount Everest.[33] Many of these have been clarified in studies that transported lowland visitors to high- or extreme-altitude locations.

- During several weeks at altitude, the gradient of O_2 between the alveoli and the blood increases during exercise, thereby increasing the arterial O_2 content and reducing hypoxia.[34]

- Chemical receptors in the carotid arteries of the neck become more sensitive to hypoxia and stimulate greater ventilation, thereby increasing O_2 levels in the blood.[35] This response is vital; without it, the brain would not sense the danger of low O_2 levels in arterial blood. Over many weeks and months, however, this adaptation is blunted. The carotid bodies in the neck may be the site of this effect because they display significant structural enlargement and biochemical changes.[35]

- Increased red blood cell production begins on the first day of altitude exposure, due to the production of the hormone erythropoietin by the kidneys.[8] However, this enhanced red blood cell production is accompanied by a decline in plasma volume, as shown in table 5.2 (page 180). The net result is that total blood volume does not increase during the first two months at altitude. After this, red cell volume continues to increase for at least one year, and probably longer.[8] It is possible that this natural polycythemia (which is primarily stimulated by trauma, parasites, or malnutrition) may be an inappropriate response at altitude, because it may reduce exercise and work performance.[4] Long-term acclimatization, however, increases plasma volume and increases red blood cell mass, thereby enlarging total blood volume.[2]

Table 5.2
Decrease in Plasma Volume During Altitude Acclimatization, in Sea Level Residents

Altitude (m)	Duration of stay (days)	Reduction in resting plasma volume (%)
3118	10	11%
3520	12	17%
3822	3-4	14%

Note: Each line of data represents a different group of subjects.

Reprinted from Young 1988.

- Bone marrow increases its iron uptake to form additional hemoglobin, beginning 48 h after exposure.[36]

- The myoglobin (muscle's counterpart to hemoglobin; an O_2 storage site) in skeletal muscle increases.[34, 37]

- Muscle fiber cross-sectional area decreases with chronic hypoxic exposure; capillary density remains unchanged or decreases; and mitochondrial density is unaltered.[31] This means that a constant capillary network supplies a smaller muscle mass, resulting in greater O_2 delivery.

- The ratio of aerobic to anaerobic metabolism increases so that energy production (i.e., ATP formation) is more efficient and the production of lactic acid is smaller.[38] This metabolic adaptation occurs, in part, because the intramuscular content of mitochondrial and respiratory enzymes increases.[39]

All of these responses serve to enhance either O_2 delivery to muscle, or increase the amount of energy generated by aerobic metabolism. This is significant because the amount of ATP that can be derived from a given amount of food is much greater for the aerobic pathway than for anaerobic glycolysis. These adaptations act harmoniously to offset the detrimental effects of hypoxia and enhance physical performance at altitude, making it more like exercise performance at sea level.

Because hypoxia limits exercise performance and induces illness at extreme altitude, it is reasonable to ask, "Is there an altitude above which humans can no longer acclimatize?"[8] The answer to this question is yes. In general, acclimatization stops and physical well-being deteriorates above 5200 m (17,060 ft); this process consists of body weight loss and deterioration of vision, memory, sleep, and computational skills.[40] There are a number of mountain peaks above this elevation in North America, including Mount McKinley in Alaska and Mount Logan in Canada. Climbers preparing to ascend the highest mountain peaks will not gain, and will likely lose, acclimatization as they spend more time above 5200 m.[8] Climbers know that acclimatization at intermediate altitude improves exercise tolerance at much higher altitudes, and shortens the time required to adjust satisfactorily. This explains why most extreme altitude expeditions ascend in stages, sometimes staying at intermediate elevations for several days or weeks.[5]

A novel technique has been used by experienced climbers since the mid-1970s.[14] A few world-class mountaineers have made fantastically rapid climbs, at very great altitudes, without supplemental O_2 or serious consequences. This is known as "rushing" a mountain. These climbers usually spent weeks working or climbing around 4300-5200 m (14,108-17,060 ft). Thus, they were acclimatized enough to protect them during the 24-60 h of ascent and descent to their base camp. It is reasonable to draw two conclusions from their experiences: (a) rushing requires sufficient acclimatization to a lower altitude before starting, and (b) rushing a high peak is possible without acclimatization if the climbing party can get up and down before illness develops.

Lowlanders Versus High-Altitude Natives

You will recall that individuals who stay at altitude for several days or weeks are named *visitors*; those who live there for months or years are *residents*. Persons who were born at altitude are known as *natives*, and those who live permanently at sea level or low altitudes are named *lowlanders*.[15] This section describes investigations that were designed to compare lowland visitors versus high-altitude natives (HAN) in controlled field and laboratory experiments.

Although relatively few studies of this type have been conducted, they add interesting bits of information to the data described in the previous section. The similarities and differences between visitors and HAN may be summarized as follows:

• HAN exhibit an exercise economy that is similar to that of lowlanders (i.e., they utilize the same amount of O_2 for a given amount of work).[31]

• Within a given ethnic group, HAN and lowlanders have similar body compositions.[31]

- High-altitude natives tend to be more polycythemic (see page 169), have larger heart dimensions, and a greater cardiac muscle mass than visitors.[41] HAN also have a more extensive coronary artery circulation (i.e., allowing greater protection against hypoxia) than lowlanders; this may explain why HAN also have a lower incidence of coronary heart disease than lowlanders.[42]

- Humans who live at high altitude during childhood appear to receive maximal benefits from acclimatization. This likely is due to the fact that the body optimizes O_2 transport mechanisms during puberty and other critical periods of growth and development.[2]

- When lowlanders become residents at an altitude of approximately 4000 m (13,123 ft), their cardiac output falls and remains depressed for several months.[41] Their reduced stroke volume is not fully compensated by increased heart rate,[43] perhaps because resting heart rate returns to nearly that seen at sea level.[2]

- The resting **basal metabolic rate (BMR)** of lowlanders was found to be 10% higher after they resided at an altitude of 5834 m for 82-113 days.[44] This enhanced BMR has been verified by other studies and is likely due to increases in the hormone thyroxin, produced by the thyroid gland.[30]

- Carl Maresh (University of Connecticut) and his former colleagues at the University of Wyoming demonstrated a greater stress hormone (e.g., cortisol) increase due to hypobaric hypoxia in lowlanders versus HAN, suggesting some acclimatization of the hypothalamic-pituitary-adrenal system to long-term hypoxia in HAN.[45] In fact, plasma cortisol levels are correlated with the duration of exposure, and increase as the symptoms of acute mountain sickness develop.[31] Cortisol levels eventually return to preexposure levels after about one week of exposure, reflecting an overall lessening (due to acclimatization) of the effects of environmental stress.[26]

In summary, HAN apparently differ from lowlanders in the development of coronary artery circulation, resting cardiovascular dynamics, and cortisol responses to stress. But these two groups have similar body compositions and exercise efficiencies.

Loss of Acclimatization

The term **deacclimatization** refers to the loss of adaptations that had been gained during hypobaric-hypoxic exposures. The rate of disappearance is different for each adaptation and each person. For example, a few days at sea level may be sufficient to render a person susceptible to altitude illness, especially high-altitude pulmonary edema (see page 186). Also, the red blood cell volume of high-altitude natives traveling to sea level decreases considerably in as little as 10 days.[46] In contrast, the improved ability to perform physical work at high altitude may persist for weeks.[2] Adaptations such as improved cardiovascular function and thermoregulation may persist for several months after descent from altitude.[41]

MEDICAL CONSIDERATIONS: HIGH ALTITUDE

A hypobaric-hypoxic environment presents the human body with a unique stressor that is found in no other environment on earth. This stressor is air containing a low oxygen content, due to reduced barometric pressure. Successful altitude acclimatization allows an individual to adapt to hypoxia, primarily by increasing O_2 delivery to cells.[47]

Hypoxia-induced medical problems at altitude are related to acclimatization. If altitude acclimatization is complete and the body successfully compensates for the hypoxia, no or few medical problems develop.[47] Incomplete acclimatization may result in various physiological states and clinical conditions that are known collectively as the altitude illnesses.

The risk of altitude illness rises with an increased *speed of ascent, altitude reached, and length of stay.* Other critical variables include age, gender, health status, previous experience at altitude, and genetic inheritance. Some factors may or may not be critical in the onset of medical problems, including diet, dehydration, infections, and emotional state.[15]

Table 5.3 presents an overview of three altitude-related medical conditions. The following paragraphs discuss these and others in detail. If you require a more thorough coverage, you may consult book chapters written by Hackett and Roach;[2] Houston;[15] Ward, Milledge, and West;[48] and Malconian and Rock.[47]

Acute Mountain Sickness (AMS)

Acute mountain sickness (AMS) involves a complex of symptoms that occur most often in unacclimatized sea-level residents who ascend rapidly to high altitude. The clinical symptoms of AMS (table 5.3) can be very debilitating. They usually develop 6-12 h after ascent, reach peak in intensity in 24-48 h, and resolve in 3-7 days as

Table 5.3
Three Altitude Illnesses[2, 16]

Condition	Symptoms
Acute mountain sickness (AMS)	Severe headache, fatigue, irritability, nausea, vomiting, loss of appetite, indigestion, flatulence, constipation, decreased urine output with normal hydration, sleep disturbance
High-altitude cerebral edema (HACE)	Loss of consciousness, staggering gait, upper body discoordination, severe weakness/fatigue, confusion, impaired mental processing, coma, drowsiness, ashen skin color
High-altitude pulmonary edema (HAPE)	Excessively rapid breathing and heart rate, rales,* breathlessness, cough producing pink frothy sputum, bluish skin color due to low blood O_2 content

* Clicking sounds heard through a stethoscope placed on the chest; this is evidense of excess mucus in the lungs.

acclimatization takes place. A few individuals experience symptoms at elevations as low as 2400 m (7874 ft), but AMS is more common over 3000 m (9842 ft). Virtually everyone will experience some symptoms if they proceed over 4300 m (14,108 ft) rapidly. Acclimatization at an intermediate altitude decreases the incidence of AMS, but even well-acclimatized persons usually develop a headache after reaching 5500 m (18,045 ft).[47]

Although the basic cause of AMS is probably related to brain swelling subsequent to hypobaric hypoxia, the symptoms are not reversed by supplemental O_2 inhalation. In moderate to severe cases of AMS, fluid retention and redistribution lead to increased intracranial pressure and lung swelling, which impairs O_2 and CO_2 exchange. One cause of this fluid retention is that renal handling of water switches from net loss or no change to a net gain of water. Concurrently, fluid is moved selectively into the intracellular space.[2]

Severe AMS gives a climber a sense of "overwhelming oppression" in which trifling work is fatiguing. The common symptom of headache can progress to cruel intensity. Apathy may be interrupted by outbursts of irritability. Eventually, one can become completely ineffective.

Treatment of AMS consists of descent, supplemental O_2, and the prescription drug acetazolamide (Diamox). Descent should continue until symptoms improve. Often a descent as small as 300 m (984 ft) will be sufficient.

Physicians, experienced climbers, and physiologists recommend several techniques to prevent AMS. These include instructions regarding graded ascent, diet, exercise, and medication.[2] *Graded ascent* refers to climbing in stages. Two to three nights should be spent at 2500-3000 m (8202-9842 ft) before going higher, and an extra night should be added for each additional 600-900 m (1968-2953 ft) attempted. Abrupt increases of more than 600 m in sleeping altitude should be avoided when at 2500 m or higher. Acclimatization may be aided by taking day trips to higher altitude, if sleep is taken at a lower elevation. This advice is summarized by the widely used phrase, "climb high, sleep low." A diet rich in carbohydrates (>70% of total calories) has been shown to reduce AMS symptoms by 30%, and increase the O_2 level of arterial blood, in a group of soldiers taken quickly to 4300 m (14,108 ft).[49, 50] Overexertion is believed to contribute to the illness, perhaps by adding to the total stress on the body. Mild exercise, though, seems to aid acclimatization.

Acetazolamide also is effective as a preventive measure for AMS. This medication aids respiratory acclimatization, prevents periods of extreme hypoxia, and maintains higher blood O_2 levels during sleep (i.e., when periodic breathing is a problem; see the section titled "Sleep," page 173).[14] Importantly, acetazolamide also counteracts fluid retention because it is a diuretic. Numerous studies indicate that acetazolamide is about 75% effective in preventing AMS in visitors to altitudes of 4000-4500 m (13,123-14,764 ft).[2] As a member of the sulfa drug family, acetazolamide carries precautions about hypersensitivity for individuals who are allergic to sulfa drugs. The drug dexamethasone offers alternative preventive properties, but it is less effective than acetazolamide and has several unwanted side effects.[2] In combination, however, acetazolamide plus a low dose of dexamethasone appears to be more effective than acetazolamide alone to ameliorate the symptoms of AMS.[51]

High-Altitude Cerebral Edema (HACE)

As mentioned in the previous section, AMS is related to brain swelling. **High-altitude cerebral edema (HACE)** appears to be an extreme form of AMS that includes loss of consciousness. The distinction between these two altitude illnesses is not clear.[2] Progression from mild AMS to unconsciousness may occur in as little as 12 h, but usually requires 3 days. HACE can be fatal if left untreated.[47] The incidence of

HACE is small, occurring in about 1% of all individuals exposed to hypobaric hypoxia. In most cases, HACE occurs with high-altitude pulmonary edema (see below).

Elevated pressures within the cranium and spine are caused by fluid shifts; these can be diagnosed with CT scan and MRI imaging techniques. The most common symptoms include a change in consciousness and lack of coordination in large muscle groups (ataxia); these characteristics may be used to differentiate HACE from AMS (see table 5.3).[2, 47] At the first sign of these two characteristics, descent should begin and dexamethasone and supplemental O_2 should be administered.[2] It is common for complications to last for weeks. Prevention of HACE is identical to that of AMS (see above).

High-Altitude Pulmonary Edema (HAPE)

High-altitude pulmonary edema (HAPE) occurs in unacclimatized lowlanders who visit high altitude by ascending rapidly. Young, healthy, active males seem to be especially susceptible, within 12 to 96 h of rapid ascent.[47] Like HACE, HAPE can be desperate, dramatic, and fatal within 12 hours if left untreated.[15] The symptoms of HAPE appear in table 5.3. Frequently, the symptoms of AMS also are present (especially nausea and headache). Mild cases of HAPE may be treated with bed rest and supplemental oxygen on site. Severe cases require descent, which is very effective, and medications.[15, 47]

Medical observations have shown that most individuals accumulate a small amount of excess fluid in the walls of alveoli when going to altitude. In some, the fluid accumulates further, leaks into the alveoli, and causes HAPE to become clinically obvious. A small percentage of humans are very susceptible to HAPE and may have up to ten attacks over the course of their lives, but no one knows why this occurs.[15]

Measures that prevent HAPE include the following: *ascending slowly, climbing and sleeping at low altitudes, limiting physical activity, avoiding cold exposure, and making visits as brief as possible.* In persons with a history of HAPE, acetazolamide may be useful.[47]

Uncommon Altitude Illnesses

You should be aware that other altitude-related medical conditions exist.[15] The first, **chronic mountain sickness (CMS)**, develops in a healthy resident after months and years at altitude. CMS is characterized by polycythemia, lethargy, weakness, sleep disturbance, bluish skin coloring (cyanosis), and change of mental status. Hypoventilation is present and is assumed to be the cause. Recovery occurs slowly after descent, and CMS may lead to disability or death.

High-altitude retinal hemorrhages (HARH) are the result of an uncommon injury to the retina of the eye. HARH are of little consequence unless they appear in certain eye structures (e.g., the macula). The cause is likely the same as AMS and HACE. The incidence of HARH is directly related to altitude; unknown below 3000 m (9842 ft), virtually all climbers experience this injury above 6700 m (21,982 ft). HARH usually go unrecognized and there is no specific treatment or prevention because they resolve spontaneously within one to two weeks. The exact cause is unknown, but most authorities believe that blood pressure surges during exercise cause ruptures in retinal capillaries.

Although there is little data available, preexisting medical problems are of great concern to many who journey to altitude.[47] For example, patients with *pulmonary disease* may experience high blood pressure, impaired O_2 and CO_2 diffusion, and exaggerated hypoxia. The risk of altitude exposure for *coronary artery disease* patients is unknown. However, it is recognized that hypoxia can induce spasms in the arteries that feed the heart, and these individuals may be at increased risk. It is clear that patients with *congestive heart failure* should avoid altitude exposure (above 2000 m, or 6562 ft) because they are likely to retain fluid and develop serious hypoxia.

RECOMMENDATIONS FOR A SAFE VISIT TO ALTITUDE

While you should remember that some altitude-caused discomfort is normal, there are certain precautions you can take to ensure the most positive experience possible. Proper nutrition and adequate fluid intake can make a big impact on your stay at altitude.

Optimal Nutrition

Because loss of appetite and strenuous exercise are obvious features of life at high and extreme altitudes, adequate caloric intake and the proper selection of nutrients is essential. This is especially appreciated by dieticians and exercise physiologists, who advise athletes about the optimal mixture of dietary carbohydrate and fat. It also is very important to the health of mountain climbers, who usually experience a marked weight loss during visits to high altitude (see above section

Spencer Swanger/Tom Stack & Associates

Proper nutrition and adequate fluid intake will help make your visit to altitude a pleasant one.

titled "Metabolism," page 172). When fed *ad libitum*, male lowlanders can lose as much as 3 kg (6.6 lb) during the first week at altitude.[26, 52]

Climbers often ask, "What should the 'best' high-altitude diet contain for optimal performance?" It appears that carbohydrate is important at high altitude. In one field study, two groups of test subjects ate either a normal or a high-carbohydrate diet.[52] The endurance performance of the group consuming the high-carbohydrate diet was superior during vigorous exercise; also, their symptoms of acute mountain sickness were fewer. Other studies have shown that failure to supply sufficient energy as carbohydrate at high altitude can result in a loss of muscle mass and decreased endurance.[53] And, during altitude acclimatization, improvements in endurance result from delaying the depletion of muscle glycogen stores.[26] Exemplifying the wisdom of the body, appetite and food preference match the need for a continuous intake of carbohydrates. Most mountain climbers prefer the taste of a high-carbohydrate–low-fat diet at altitude, and they find fatty foods to be distasteful.[48] This, coupled with the metabolic adaptation described in the next paragraph, may explain why body weight decreases during the initial days of altitude exposure. In other words, because the diet is low in calories and low in fat, the body utilizes its own adipose fat stores as fuel.

Interestingly, biochemical experiments show that cellular utilization of nutrients is altered during the first three weeks of altitude acclimatization. Fat becomes the predominant fuel during exercise, and carbohydrate utilization decreases.[26] This defends local carbohydrate stores of muscle and liver glycogen, and serves to enhance endurance time to exhaustion. In mountain climbing at extreme altitude, this could be the difference between life and death. The mechanisms for the sparing of muscle glycogen and the shift to fat metabolism are unknown, but three theories exist. First, this phenomenon may be due to the prolonged, low-intensity exercise that is characteristic of climbing and that stimulates fat metabolism preferentially. Second, a low-calorie–low-carbohydrate diet, coupled with high energy expenditure (e.g., a day-long climb), may reduce glycogen stores in liver and muscle, forcing a shift to fat metabolism. Third, the autonomic nervous system (see chapter 1) may undergo a change. It has been observed that high-altitude exposure chronically increases sympathetic nervous stimulation and elevates plasma norepinephrine.[54] In turn, norepinephrine causes adipose tissue to release free fatty acids and glycerol into the blood. The free fatty acids then are absorbed by muscle fibers via mass action.[26]

Fluid Balance

Body water losses may be misjudged by novice climbers because the air temperature is cool on high peaks, decreasing by about 0.6°C (1.1°F) for each elevation increase of 100 m (328 ft). It is easy to forget that dehydration occurs during mountain climbing. The average daily water loss at altitude includes urination (1.3 L), feces (0.1-0.2 L), sweat (0.1 L), and water that passes through the lungs and skin (0.7-1.1 L).[53] Because the air becomes drier (and because hyperventilation is stimulated) at altitude, water loss can be large enough to cause marked dehydration over several days. Pugh, for example, found that mountain climbers required water intakes of up to 6.2 L/day because of fluid losses in urine, feces, sweat, and expired gases;[11] their actual rate of water consumption was only 1.6 to 2.5 L/day. Pugh later estimated that 3-4 L of water was required by humans climbing for 7 h to maintain a normal adult urine output of 1.4 L/day.[43] His observations were supported by the field study of Consolazio and colleagues in 1972.[55] They observed that males lost 3.5-4.0 kg (7.7-8.8 lb) during a 12-day sojourn at 4325 m (14,190 ft). Body water loss accounted for approximately 50% of the total weight loss; body fat, protein, and minerals accounted for the remainder.

Interestingly, the fluid compartments of the body may not contribute to this loss equally. Hoyt and colleagues observed this phenomenon in young adult males who visited Pikes Peak, Colorado (elevation 4325 m), for 10 days.[56] These men lost 1.5 L of total body water, extracellular water expanded by 1.2 L, plasma volume decreased by 20%, and intracellular fluid contracted by 2.7 L. A strikingly similar pattern of fluid equilibration had been reported earlier in lowlanders who accomplished a 6-day stay at the same site,[57] despite the fact that plasma volume and extracellular fluid volume usually expand and shrink in concert.

To complicate this dehydration, an altitude-induced diuresis (e.g., increased urine production) often occurs in healthy individuals during the initial days of exposure to a hypobaric-hypoxic environment.[53] This means that the ideas presented in chapter 2 regarding dehydration and fluid replacement in hot environments apply to high-altitude environments as well. Also, the urine color chart on the rear cover of this book should help avoid dehydration.

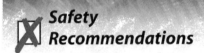

Safety Recommendations

As with all other environments described in this book, recreation and residence at high altitude offer enjoyable benefits, when attempted safely. If you plan to travel to a high- or extreme-altitude environment, the following eight suggestions will help you deal with hypobaria and hypoxia wisely.

- Ascend slowly.
- Conduct your climb in stages. If going to an elevation over 3000 m (9842 ft), limit your rate of ascent to 300 m (984 ft) per day.
- Climb with an experienced guide or team.
- Avoid dehydration and overexertion.
- Eat a high-carbohydrate diet to reduce AMS symptoms.
- Use medication as a preventive measure.
- Like diving into the hyperbaric, underwater environment described in chapter 4, mountain climbing in a hypobaric-hypoxic environment offers significant health risks. High-altitude recreation and labor should not be attempted without careful plan-

ning. Your expectations about performance at altitude should be tailored to match your climbing experience, knowledge, fitness level, and acclimatization state.

- Regarding a climber's *attitude* about altitude, one of the most influential climbers of this century (Reinhold Messner) has stated: "the best climber is not the one who does a crazy thing once or twice and dies the second time; the best climber is one who does many things on the highest level and survives."[32]

REFERENCES

1. U.S. Navy. 1985. *U.S. Navy diving manual. Air diving.* Vol. 1, 1-18. Washington, DC: U.S. Government Printing Office.

2. Hackett, P.H., & Roach, R.C. 1995. High-altitude medicine. In *Wilderness medicine,* edited by P.S. Auerbach, 1-37. St. Louis: Mosby.

3. Noble, B.J. 1986. *Physiology of exercise and sport.* St. Louis: Times Mirror/ Mosby, 426-451.

4. West, J.B. 1996. Physiology of extreme altitude. In *Handbook of physiology. Section 4: Environmental physiology.* Vol. II, edited by M.J. Fregly & C.M. Blatteis, 1307-1325. New York: Oxford University Press.

5. Miller, A.T. 1974. Altitude. In *Environmental physiology,* edited by N.B. Slonim, 350-375. St. Louis: Mosby.

6. Reeves, J.T.; Groves, B.M.; Sutton, J.R.; Wagner, P.D.; Cymerman, A.; Malconian, M.K.; Rock, P.B.; Young, P.M.; & Houston, C.S. 1987. Operation Everest II: Preservation of cardiac function at extreme altitude. *Journal of Applied Physiology: Respiratory, Environmental, Exercise Physiology 63*: 531-539.

7. Sarnquist, F.H.; Schoene, R.B.; Hackett, P.H.; & Townes, B.D. 1986. Hemodilution of polycythemic mountaineers: Effects on exercise and mental function. *Aviation, Space, and Environmental Medicine 57*: 313-317.

8. Haymes, E.M., & Wells, C.L. 1986. *Environment and human performance.* Champaign, IL: Human Kinetics, 69-91.

9. Fuentes, R.J.; Rosenberg, J.M.; & Davis, A. 1996. *Athletic drug reference '96.* Research Triangle, NC: Glaxo Wellcome, 37-40.

10. Ryn, Z. 1971. Psychopathology in alpinism. *Acta Medica Polinica 12*: 453-467.

11. Pugh, L.G.C.E. 1962. Physiological and medical aspects of the Himalayan Scientific and Mountaineering Expedition, 1960-1961. *British Medical Journal 2*: 621-633.

12. Anholm, J.D. 1992. Operation Everest II: Arterial oxygen saturation and sleep at extreme simulated altitude. *American Review of Respiratory Disease 145*: 817-826.

13. Reite, M.; Jackson, D.; Cahoon, R.L.; & Weil, J.V. 1975. Sleep physiology at high altitude. *Electroencephalography and Clinical Neurophysiology 38*: 463-471.

14. Hackett, P.H. 1987. Respiratory stimulants and sleep periodic breathing at high altitude: Almitrine versus acetazolamide. *American Review of Respiratory Diseases 135*: 896-898.

15. Houston, C.S. 1984. Man at altitude. In *Sports Medicine,* edited by R.H. Strauss, 344-360. Philadelphia: Saunders.

16. Messner, R. 1979. *Everest: Expedition to the ultimate.* London: Kaye and Ward.

17. Craig, A.B. 1969. Olympic 1968: A post mortem. *Medicine & Science in Sports 1*: 177-180.

18. Green, H.J.; Sutton, J.; Young, A.; Cymerman, A.; & Houston, C.S. 1989. Operation Everest II: Muscle energetics during maximal exhaustive exercise. *Journal of Applied Physiology: Respiratory, Environmental, Exercise Physiology 66*: 142-150.

19. Åstrand, P.O., & Rodahl, K. 1977. *Textbook of work physiology.* New York: McGraw-Hill.

20. Adams, W.C.; Bernauer, E.M.; Dill, D.B.; & Bomar, J.B. 1975. Effects of equivalent sea-level and altitude training on VO_2max and running performance. *Journal of Applied Physiology 39*: 262-266.

21. Maher, J.T.; Jones, L.G.; & Hartley, L.H. 1974. Effects of high-altitude exposure on submaximal endurance capacity of men. *Journal of Applied Physiology 37*: 895-898.

22. Horstman, D.; Weiskopf, R.; & Jackson, R.E. 1980. Work capacity during a 3-week sojourn at 4,300 m: Effects of relative polycythemia. *Journal of Applied Physiology: Respiratory, Environmental, Exercise Physiology 49*: 311-318.

23. Hoyt, R.W., & Honig, A. 1996. Body fluid and energy metabolism at high altitude. In *Handbook of physiology.* Section 4: *Environmental physiology.* Vol. II, edited by M.J. Fregly & C.M. Blatteis, 1277-1289. New York: Oxford University Press.

24. Buskirk, E.R.; Kollias, J.; Akers, R.F.; Prokop, E.K.; & Picon-Reatigue, E.P. 1967. Maximal performance at altitude and on return from altitude in conditioned runners. *Journal of Applied Physiology 23*: 259-266.

25. Young, A.J. 1988. Human acclimatization to high terrestrial altitude. In *Human performance physiology and environmental medicine at terrestrial extremes,* edited by K.B. Pandolf, M.N. Sawka, & R.R. Gonzalez, 497-543. Indianapolis: Benchmark Press.

26. Bannister, E.W., & Woo, W. 1978. Effects of simulated altitude training on aerobic and anaerobic power. *European Journal of Applied Physiology 38*: 55-69.

27. Balke, B.; Daniels, J.T.; & Faulkner, J.A. 1967. Training for maximum performance at high altitude. In *Exercise at altitude*, edited by R. Margaria, 179-186. New York: Excerpta Medical Foundation.

28. Leadbetter, G. 1992. The effect of intermittent altitude exposure on acute mountain sickness. Abstract. *Annual meeting of the Southwest Chapter of American College of Sports Medicine.* San Diego, CA.

29. Levine, B.D., & Stray-Gundersen, J. 1992. A practical approach to altitude training: Where to live and train for optimal performance enhancement. *International Journal of Sports Medicine 13, Supplement 1*: S209.

30. Levine, B.D., & Stray-Gundersen, J. 1997. "Living high-training low": Effect of moderate-altitude acclimatization with low-altitude training on performance. *Journal of Applied Physiology 83*: 102-112.

31. Cerretelli, P., & Hoppeler, H. 1996. Morphologic and metabolic response to chronic hypoxia: The muscle system. In *Handbook of physiology*. Section 4: *Environmental physiology*. Vol. II, edited by M.J. Fregly & C.M. Blatteis, 1155-1181. New York: Oxford University Press.

32. O'Connell, N. 1993. *Beyond risk. Conversations with climbers.* Seattle: The Mountaineers, 155.

33. Fulco, C.S. 1988. Human acclimatization and physical performance at high altitude. *Journal of Applied Sport Science Research 2*: 79-84.

34. Dempsey, J.A.; Reddan, W.G.; Birnbaum, M.L.; Foster, H.V.; Thoden, J.S.; Grover, R.F.; & Rankin, J. 1971. Effects of acute through life-long hypoxic exposure on exercise pulmonary gas exchange. *Respiratory Physiology 13*: 62-89.

35. Bisgard, G.E., & Forster, H.V. 1996. Ventilatory responses to acute and chronic hypoxia. In *Handbook of physiology*. Section 4: *Environmental physiology*. Vol. II, edited by M.J. Fregly & C.M. Blatteis, 1207-1239. New York: Oxford University Press.

36. Reynafarje, C.; Lozano, R.; & Valdiviesto, J. 1959. The polycythemia of high altitudes: Iron metabolism and related aspects. *Blood 14*: 433-455.

37. Frisancho, A.R. 1975. Functional adaptation to high altitude hypoxia. *Science 187*: 313-319.

38. Hockachka, P.W. 1996. Metabolic defense adaptations to hypobaric hypoxia in man. In *Handbook of physiology*. Section 4: *Environmental physiology*. Vol. II, edited by M.J. Fregly & C.M. Blatteis, 1115-1123. New York: Oxford University Press.

39. Rumsey, W.L., & Wilson, D.F. 1996. Tissue capacity for mitochondrial oxidative phosphorylation and its adaptation to stress. In *Handbook of physiology*. Section 4: *Environmental physiology*. Vol. II, edited by M.J. Fregly & C.M. Blatteis, 1095-1113. New York: Oxford University Press.

40. Pugh, L.G.C.E. 1964. Animals in high altitudes: Man above 5,000 meters—Mountain exploration. In *Handbook of physiology*. Section 4, edited by D.B. Dill, E.F. Adolph, & C.G. Wilber, 861-888. Washington, DC: American Physiological Society.

41. Mirrakhimov, M.M., & Winslow, R.M. 1996. The cardiovascular system at high altitude. In *Handbook of physiology*. Section 4: *Environmental physiology*. Vol. II, edited by M.J. Fregly & C.M. Blatteis, 1241-1257. New York: Oxford University Press.

42. Arias-Stella, J., & Topilsky, M. 1971. Anatomy of the coronary circulation at high altitude. In *High altitude physiology: Cardiac and respiratory aspects,* edited by R. Porter & J. Knight, 149-154. Edinburgh: Churchill Livingstone.

43. Pugh, L.G.C.E. 1964. Cardiac output in muscular exercise at 5800 m (19,000 ft). *Journal of Applied Physiology 19*: 441-447.

44. Gill, M.B., & Pugh, L.G.C.E. 1964. Basal metabolism and respiration in men living at 5800 m (19,000 ft). *Journal of Applied Physiology 19*: 949-954.

45. Maresh, C.M.; Nobel, B.J.; Robertson, K.L.; & Harvey, J.S. 1985. Aldosterone, cortisol, and electrolyte responses to hypobaric hypoxia in moderate-altitude natives. *Aviation, Space, and Environmental Medicine 56*: 1078-1084.

46. Clark, C.F.; Heaton, R.K.; & Wrens, A.N. 1983. Neuropsychological functioning after prolonged high altitude exposure in mountaineering. *Aviation, Space, and Environmental Medicine 54*: 202-207.

47. Malconian, M.K., & Rock, P.B. 1988. Medical problems related to altitude. In *Human performance physiology and environmental medicine at terrestrial extremes,* edited by K.B. Pandolf, M.N. Sawka, & R.R. Gonzalez, 545-563. Indianapolis: Benchmark Press.

48. Ward, M.P.; Milledge, J.S.; & West, J.B. 1989. *High altitude medicine and physiology.* Philadelphia: University of Pennsylvania Press, 283-292.

49. Consolazio, C.F.; Matoush, L.O.; Johnson, H.L.; Krzywicki, H.J.; Daws, T.A.; & Isaac, G.J. 1969. Effects of a high-carbohydrate diet on performance and clinical symptomatology after rapid ascent to high altitude. *Federation Proceedings 28*: 937-943.

50. Hansen, J.E.; Hartley, L.H.; & Hogan, R.P. 1972. Arterial oxygen increased by high-carbohydrate diet at altitude. *Journal of Applied Physiology 33*: 441-445.

51. Bernhard, W.N.; Schalick, R.N.; Delaney, P.A.; Bernhard, T.M.; & Barnas, G.M. 1998. Acetazolamide plus low-dose dexamethasone is better than acetazolamide alone to ameliorate symptoms of acute mountain sickness. *Aviation, Space, and Environmental Medicine 69*: 883-886.

52. Consolazio, C.F.; Matoush, L.O.; Johnson, H.L.; & Daws, T.A. 1968. Protein and water balances of young adults during prolonged exposure to high altitude (4300 m). *American Journal of Clinical Nutrition 21*: 154-161.

53. Cymerman, A. 1996. The physiology of high-altitude exposure. In *Nutritional needs in cold and in high-altitude environments,* edited by B. Marriott & S.J. Carlson, 295-317. Washington, DC: National Academy Press.

54. Cunningham, W.L.; Becker, E.J.; & Kreuzer, F. 1965. Catecholamines in plasma and urine at high altitude. *Journal of Applied Physiology 20*: 607-610.

55. Consolazio, C.F.; Johnson, H.L.; & Krzywicki, H.J. 1972. Body fluids, body composition, and metabolic aspect of high-altitude adaptation. In *Physiological adaptations: Desert and mountain,* edited by M.K. Yousef, S.M. Horvath, & R.W. Bullard. New York: Academic Press.

56. Hoyt, R.W.; Durkot, M.J.; Kamimori, G.H.; Schoeller, D.A.; & Cymerman, A. 1992. Chronic altitude exposure (4300 m) decreases intracellular and total

body water in humans. Abstract. In *Hypoxia and mountain medicine,* edited by J.R. Sutton, G. Coates, & C.S. Houston, 306. Burlington, VT: Queen City.

57. Krzywicki, H.J.; Consolazio, C.F.; Johnson, H.L.; Nielsen, W.C.; & Barnhart, P.A. 1971. Water metabolism in humans during acute high-altitude exposure (4300 m). *Journal of Applied Physiology 30*: 806-809.

chapter 6
Air Pollution: Exercise in the City

*From the standpoint of coaches and athletes, carbon monoxide is the most important of the primary pollutants. The principle source of carbon monoxide is automotive exhaust. Carbon monoxide exerts its effect by binding to and blocking the oxygen-binding sites on **hemoglobin** in the red blood cells [forming the molecule HbCO]. Hemoglobin has an affinity for carbon monoxide that is 230 times greater than its affinity for oxygen.... This means that increased levels of carbon monoxide in the blood compromise both the transport of oxygen in the blood, and the extraction of oxygen to the tissues.... The immediate impact of this on exercise performance is that as the concentration of HbCO in the blood increases, there is a decrement in maximum oxygen consumption ... [and a decrease] in maximal exercise time.*

—Peter N. Frykman, 1988[1]

Airborne contaminants represent environmental stressors that primarily affect the respiratory and cardiovascular systems. However, air pollutants also may cause inflammation and may increase the levels of blood-borne immune factors. It is not valid to categorize the effects of carbon monoxide, ozone, sulfur dioxide, nitrogen dioxide, and peroxyacetyl nitrate as one environmental stressor because they each affect health and performance via unique mechanisms and to a different degree. Although we cannot selectively filter air pollutants, this chapter presents techniques to reduce the effects of contaminants on exercise performance and health. High-risk individuals, such as those with asthma, also are identified.

POLLUTION: A MODERN ENVIRONMENTAL STRESSOR

As described in the preface of this book, earth's environment can be extreme in many ways. Environmental features, including temperature, pressure, wind speed, and moisture, represent stressors that challenge survival and diminish performance. Although adequate adaptive responses occur (acclimatization) in response to repeated exposures to different stressors, one aspect of our environment seems to have outpaced our hereditary capabilities: air pollution. The ocean of air that surrounds us has changed so rapidly in the last century, especially due to the internal combustion engine, that it presents a new challenge to our adaptive potentials.[2] Because research involving individual pollutants is in its infancy, we are still learning simple facts regarding the impact of air quality on exercise and work performance. The details below verify that the effects of each pollutant are distinctive, and that each should be viewed as a unique stressor that the body may, or may not, be genetically equipped to deal with.

In contrast to heat and humidity, which affect virtually all organs in the body, air pollutants affect few body systems directly (figure 6.1). Despite this difference, weather fronts that contain heat and humidity may act to magnify the effects of air pollution on health and physical performance. Figure 6.2 (page 200) illustrates this connection.[3] Panel A shows normal atmospheric conditions, in which the air above the earth gets colder as altitude increases. This decrease in temperature amounts to a decrease of approximately 0.6°C (1.1°F) for every 100 m (328 ft) increase in altitude. In these conditions, the pollutants, which are relatively warm, rise through the layers of denser, cold air. Panel B shows the conditions of a thermal inversion, during which a layer of warm or hot air is "sandwiched" between two layers of colder air. Such a warm air layer may be very stable and persist for several days, until a different weather front arrives and the thermal inversion is disrupted. The warm air layer forms a ceiling that traps the

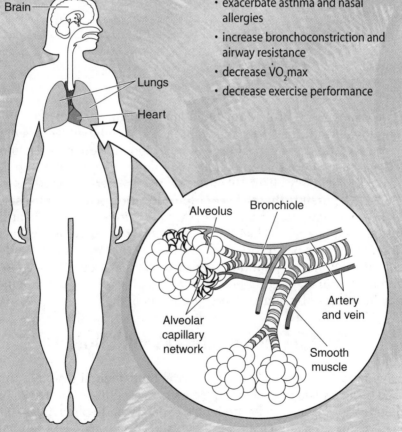

The Body's Responses to Stress

Nervous System

- brain senses irritation due to the presence of a pollutant
- brain sends efferent output that may/may not affect homeostasis

Cardiovascular System

- pollutants may decrease blood O_2-carrying capacity

Immune System

- pollutants may cause inflammation and increase blood-borne immune factors

Respiratory System

Pollutants may:

- increase respiratory rate
- reduce pulmonary function
- exacerbate asthma and nasal allergies
- increase bronchoconstriction and airway resistance
- decrease $\dot{V}O_2$max
- decrease exercise performance

Brain

Lungs

Heart

Alveolus

Bronchiole

Artery and vein

Alveolar capillary network

Smooth muscle

Figure 6.1 Air pollution affects the lungs, airways, and blood contents.

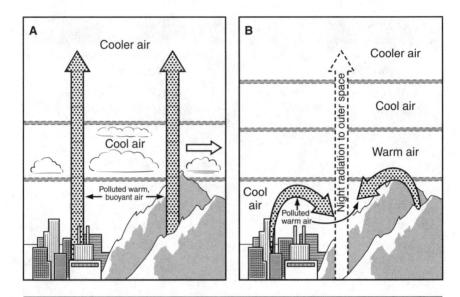

Figure 6.2 Normal atmospheric conditions (A) and a thermal inversion (B).
Reprinted from Frykman 1988.

upward movement of airborne pollutants, causing them to move horizontally beneath the thermal inversion. The combined factors of an environmental trap and a source of pollution are common to almost all locations where there is an accumulation of atmospheric pollution that negatively affects health and exercise performance.[1]

It even is possible that the geographic location of a city causes it to be an environmental trap. Denver, Colorado, for example, sits at the bottom of a large bowl-like basin on the eastern edge of the Rocky Mountains. This basin forms an ideal collecting trap for the industrial and automotive pollutants generated in that area.[4] Thus, anyone who lives, works, or competes in a geographic area such as Denver may be affected by atmospheric contaminants.

TESTS OF PULMONARY FUNCTION

Pulmonary function tests (PFT) are the fundamental research tools used by scientists to evaluate the effects of air pollutants. These PFT include measurements of the lung's capacity and volume, gas diffusion, airway resistance to gas flow, and forced exhalation-inhalation.[4, 6, 7] The specific tests that often demonstrate reduced pulmonary function include the following:

- *Forced expiratory volume (FEV$_t$):* the volume of air exhaled by a maximal effort during a specific time period (e.g., 1 sec, FEV$_1$)
- *Forced vital capacity (FVC)*: the volume of air that can be expelled from the lungs during a maximal effort following a maximal inspiration
- *Forced midexpiratory flow (FEF)*: the average rate of airflow over the middle half of the FVC
- *Airway resistance (R$_{aw}$)*: the difference between the pressure in the alveoli and the mouth, divided by the rate of air flow
- *Diffusing capacity of the lung (D$_L$)*: a measure of the lung's ability to move a gas from the alveoli into the capillary bloodstream; carbon monoxide is the gas most commonly used to measure D$_L$

AIR POLLUTION AND PHYSICAL PERFORMANCE

Air pollution, like air itself, is made up of many different gases. These various gases can take a heavy toll on your physical performance.

Carbon Monoxide

Airborne pollutants are categorized as primary pollutants and secondary pollutants. **Primary pollutants** are those that exert their physiological influence directly from the source of pollution (e.g., carbon monoxide, nitrogen oxides, sulfur oxides, and particulates). **Secondary pollutants** are formed by the interaction of primary pollutants with other compounds, ultraviolet light, or with each other. They include aerosols, ozone, and peroxyacetyl nitrate.[6]

Physiologically, **carbon monoxide** is the most significant primary pollutant because it alters the ability of red blood cells (erythrocytes) to carry oxygen to skeletal muscle and other tissues. Found mostly in automobile exhaust, carbon monoxide (chemical formula CO) attaches to and blocks the oxygen-binding sites on the molecule hemoglobin, which is a complex protein that is located inside erythrocytes. Hemoglobin (Hb) combines readily with oxygen (O$_2$) to form HbO$_2$ or with CO to form **carboxyhemoglobin (HbCO)**. However, the affinity of Hb for CO is about 230 times greater than its affinity for O$_2$, indicating that CO has great potential to alter oxygen transport in blood.

The complex psychological, physiological, and clinical effects of different levels of HbCO are summarized in table 6.1 (page 203).[1] Column 3 of this table presents the percent carboxyhemoglobin in blood,

Carbon monoxide hampers the ability of red blood cells to carry oxygen and limits the performance of athletes in urban settings.

ranging from the resting normal level for nonsmokers (<1% HbCO) to levels that induce coma, respiratory failure, and death (60-80% HbCO). Columns 1 and 2 describe the environmental levels (expressed as parts per million, ppm) of CO that are required to produce the blood levels of HbCO in column 3, after 1 h and 8 h of resting exposure, respectively. Column 4 presents the observable effects that begin at each of the blood values in column 3. Excessively high levels of HbCO in blood (i.e., >60%), as occur in accidental CO poisoning or suicide attempts, are associated with extensive destruction of cardiac muscle tissue and nerve cells. If an individual survives such an exposure, defective central nervous system performance and myocardial damage result; the extent of such injury is related to the degree of hypoxia (oxygen scarcity) experienced.[2] One report, in fact, observed that heavy exercise at low ambient CO levels could induce states of CO poisoning due to increased ventilation.[8]

In terms of exercise, inhalation of CO was first shown to limit physical performance in 1941, when Chiodi and colleagues demonstrated that high blood concentrations of 40-50% HbCO made it impossible for test subjects to perform routine laboratory tasks at low exercise intensities.[9] In fact, some subjects collapsed during these experiments. Horvath reviewed six other studies in which humans performed light

Table 6.1
Effects of Different Levels of Carboxyhemoglobin (HbCO) on Psychological, Physiological, and Clinical Factors

Approximate ambient level of CO (ppm)			
1 h exposure	8 h exposure	%HbCO in blood	Threshold effects
0	0	0.3–0.7	Physiologic norm for nonsmokers
55–80	15–18	2.5–3.0	Cardiac function decrements occur in impaired individuals; blood flow alterations; increased red blood cell production after long exposure
110–170	30–45	4–6	Visual impairment, vigilance decrement, reduced maximal work capacity
280–575	75–155	10–20	Slight headache, lassitude, dilation of skin blood vessels; coordination problems
575–860	155–235	20–30	Severe headache, nausea
860–1155	235–310	30–40	Muscular incoordination, nausea, vomiting, dimness of vision
1430–1710	390–470	50–60	Syncope, convulsion, coma
1710–2000	470–550	60–70	Coma, depressed cardiac and respiratory function, sometimes fatal, permanent defect
2000–2280	550–630	70–80	Respiratory failure, death

Abbreviation: %HbCO, percent carboxyhemoglobin.

Adapted from Horvath 1981.

to moderate exercise tasks with low blood HbCO levels, ranging from 10.7 to 20.1%.[2] These studies uniformly showed no change in submaximal oxygen uptake, despite the presence of HbCO. The only observable physiological sign of increased physiological strain was a slightly higher heart rate during exercise, and increased **pulmonary ventilation ($\dot{V}_E$)** at higher submaximal exercise intensities (70% $\dot{V}O_2$max).[6] Thus, it appears that ambient CO levels resulting in HbCO levels below 15% have little impact on physiological responses to low-intensity exercise (i.e., 35-70% $\dot{V}O_2$max) lasting 5-60 min.[2, 6] Also, it is important to recognize that the HbCO levels reported in these studies are well *above* those normally experienced when air quality is poor.[7] In 1972, for example, a 90 min ride on the Los Angeles freeway in an open automobile resulted in blood HbCO levels of 5.1%.[10]

In contrast, maximal exercise performance (i.e., total exercise time or $\dot{V}O_2$max) appears to be inversely related to the carboxyhemoglobin concentration in blood. The minimal threshold of this effect has been shown statistically to be 4.3% HbCO for measurements of $\dot{V}O_2$max.[11] Significant decrements in maximal exercise time, leg tiredness, and perceived exertion have been reported at even lower HbCO levels (e.g., 3.3-4.3%).[11, 12] Figure 6.3 presents a graph of the HbCO content of blood plotted against the decrease in maximal aerobic power. The statistical relationship between these two variables is strong and significant (i.e., correlation coefficient of $r = 0.94$, $p < .0001$), and the equation that predicts the decrement in $\dot{V}O_2$max, when percent HbCO is known, is[11]

$$\text{percent decrease of } \dot{V}O_2\text{max} = [0.91\ (\%\ \text{HbCO})] + 2.2$$

This graph also supports the two prominent theoretical mechanisms explaining why HbCO reduces maximal exercise capacity at low blood levels, and mild to moderate exercise capacity at very high blood levels.[2] The first theoretical mechanism of carbon monoxide toxicity involves hypoxia. The presence of HbCO in the blood reduces the total availability of oxygen because CO competes with O_2 for the binding site on the hemoglobin molecule, and inhibits the release of O_2 from the hemoglobin molecule in tissue (e.g., muscle) capillary beds. The second theory recognizes that CO also binds with iron-rich compounds such as myoglobin in skeletal muscle and with cytochrome molecules in mitochondria (i.e., the site of the electron transport and oxidative phosphorylation steps of aerobic metabolism that generate ATP). This mechanism of CO toxicity involves blocking the production of the energy-rich compound ATP, which is used by cells to power virtually all of life's processes (such as muscle contraction, cell reproduction and growth, nerve conduction, digestion, transport of molecules across cell membranes), at the cellular level within the mitochondrial cytochrome system.[2] Whichever of these two mechanisms reduces maximal work capacity, it appears that the effects of CO are magnified at high altitude, where the ambient partial pressure of oxygen is lower and blood O_2 content is lower.[13] Similarly, people with cardiovascular disorders such as *angina pectoris* have been shown to have a reduced exercise time to the onset of clinical *angina* symptoms when blood levels of HbCO, induced by secondhand (passive) smoke, were only 2%.[9, 10] In comparison, smokers who inhale generally have blood HbCO levels of 4-7%,[11] as do runners who train in New York City.[14]

Clearly, CO demonstrates that air pollution can adversely influence exercise performance. However, most research has focused on the first organ to be affected—the lung—and little has focused on other organ systems or on groups besides adult males (e.g., females, youth,

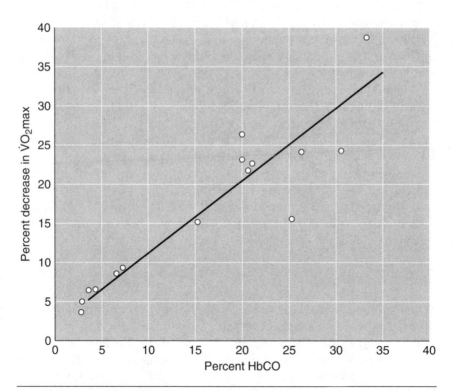

Figure 6.3 The relationship between blood carboxyhemoglobin content (percent HbCO) and decrease in maximal aerobic power (percent decrease $\dot{V}O_2$max). Reprinted from Horvath 1981.

the elderly). The effects of the remaining primary pollutants on the physiology of human performance are subtle, in comparison to those of CO.

Nitrogen Dioxide

The family of nitrogen oxide compounds (general chemical formula, NO_x) are released into the atmosphere as air pollutants during heavy motor vehicle or aircraft engine combustion, fire fighting, and cigarette smoking. Of the six types of compounds in this family, **nitrogen dioxide (NO_2)** has been tested extensively because it is known to be potentially harmful to health.[2] Individuals who have respiratory disorders, such as chronic bronchitis or obstructive pulmonary disease, are believed to be more susceptible to the adverse effects of NO_2 exposure.[6] Peak hourly levels have reached 0.34 ppm NO_2 in Los Angeles.[2]

Nitrogen dioxide is a soluble gas and can be absorbed by the mucus of the nasopharyngeal cavity, where it converts to nitrous acid and nitric acid. When inhaled in high concentrations, as occurs in agricultural or industrial accidents, pulmonary dysfunction may persist for periods of 2.5 to 13.5 years. These deficits include diminished exercise tolerance time and breathlessness during exertion.[15] In healthy humans, according to the review of Pandolf,[6] low-level NO_2 exposure (0.4-2.7 ppm) does not affect physical performance adversely during submaximal exercise; this includes athletes.[16] Observations of maximal exercise performance have not been published to date, in either healthy or patient populations, but are justified by animal studies that have shown effects of NO_2 on deep airways.[17] In fact, the primary difficulty associated with low-level NO_2 inhalation appears to be a mild irritation of the upper respiratory tract and impairment of mucociliary activity in bronchial tubes.[18]

Sulfur Dioxide

When sulfur-containing fossil fuels are burned, the sulfur oxide family of compounds (general chemical formula, SO_x) are formed; this family includes sulfuric acid, sulfate, and **sulfur dioxide (SO_2)**. This latter substance accounts for about 90% of all initial sulfurous byproducts of combustion, and is a highly soluble gas that exerts its influence mainly as an upper respiratory tract irritant.[6] When inhaled, SO_2 can produce acute bronchial tube constriction, by stimulating the vagus nerve in the pharynx region of the throat, behind the tongue. This **bronchoconstriction** results in an increased resistance to the flow of air.[19] Interestingly, this airway resistance (abbreviated R_{aw}) increases at rest, as the volume of inspired air increases.[20] Two factors are involved: (a) increased ventilation per minute ($\dot{V}_E$) augments the total volume of SO_2 delivered to sensitive bronchial sites; and (b) increased $\dot{V}_E$ decreases the percentage of air that is inspired through the nose, bypassing an important defense against SO_2, nasal "scrubbing" of soluble gases. Nasal scrubbing is a very effective process that absorbs more than 99% of SO_2 during quiet breathing.[21] Both of these factors, therefore, allow a greater volume of SO_2 into the deeper segments of the pulmonary tree and theoretically should increase bronchoconstriction.[22]

Because exercise increases $\dot{V}_E$, the foregoing information logically suggests that R_{aw} will increase if SO_2 is inhaled during labor, sport-specific competition, or athletic competition. And, if R_{aw} increases, it is theoretically reasonable that physical performance will suffer, because maximal ventilation should be reduced. Although no studies have incorporated maximal exercise performance thus far,[5] this topic is worthy of future research. At least ten studies, however, have evaluated submaximal exercise performance. These have been reviewed by various authors, who have drawn the following three conclusions about the effects of SO_2 inhalation on submaximal exercise performance in healthy adults.

- Some, not all, individuals show significant decrements in lung function following low levels of SO_2 exposure.[23]
- The threshold of effects on performance lies between 1.0 and 3.0 ppm SO_2.[2, 6]
- At air SO_2 levels of 5.0 ppm, decreases occur in expiratory flow, tracheal-bronchial clearance of mucus, and other measurements of lung function.[2, 6]

People with asthma or those with nonlocalized allergies are generally more sensitive to the bronchoconstrictive effects of SO_2 than those without.[21] Asthmatics, for example, experience worsened nasal symptoms and reduced airflow. They experience bronchospasm within minutes of inhaling air that contains as little as 0.2-0.5 ppm SO_2.[6] Nonasthmatic individuals typically require exposures of SO_2 above 2 ppm, which rarely occur in the air of large cities. Air temperature and humidity also play a critical role in the development of bronchospasm in those who have asthma. Breathing SO_2 in cold-dry air will produce a faster and more intense bronchospasm than breathing SO_2 in warm-moist air.[24]

Although no research has evaluated adaptation during exercise, various studies suggest that the human body makes favorable adaptive responses to repeated SO_2 exposures, at rest.[6] For example, industrial workers did not respond to 5 ppm SO_2 (as did normal control subjects) if they regularly were exposed to 10 ppm SO_2 at their worksite.[25] Adaptation is possible even within a single 6 h period, as shown during observations of 15 healthy males.[26] During this experiment, ambient levels of SO_2 slowly increased from 1 to 25 ppm. The test subjects tolerated the latter, high level of SO_2 very well, possibly because the concentration was increased slowly. But the investigators, who entered the environmental chamber periodically and were very disturbed by large, abrupt changes in the SO_2 level, found the discomfort of a 25 ppm exposure to be "almost intolerable."[2]

Exposures to high concentrations of SO_2, which are many times greater than typical outdoor city pollution levels, usually occur in confined industrial settings.[24] One report, published after an accidental underground explosion, tracked the recovery of nonsmoking miners for two years.[27] Within three weeks of this incident, these laborers showed severe airway obstruction, markedly reduced exercise capabilities, hypoxemia, and active inflammation. Only partial recovery was seen during this clinical study, with most improvements occurring within the first year.

Fine Particulates

Many inhaled substances contain, or are classified as, particulate matter; these include dust, acidic aerosols, tobacco smoke, wood smoke, pollen, bacteria, and sulfuric acid from sulfur-containing fuels.[24] The types of particulates found in the air of any given city depend on human activities (e.g., burning, type of fuels used) and environmental factors (e.g., wind direction and speed, thermal inversion layers). The effects of particulates on lung tissue are related to particle size, total particle mass, chemical composition, deposition patterns, oral versus nasal breathing, and the defense mechanisms of the lung.[28] Larger particles (5-10 μm) are deposited in the nasopharyngeal region and cause inflammation, congestion, and ulceration. The tracheobronchial region of the pulmonary system receives particles with a diameter of 3-5 μm; these stimulate bronchospasm, bronchial congestion, and bronchitis. Only the smallest particles (0.5-3.0 μm) reach the alveoli. **Mucociliary transport** and **phagocytosis** are employed for lung clearance after particles have been inhaled.[29] Both

A Detroit factory worker wears a respirator to prevent the inhalation of fine particulates.

asthmatic and nonasthmatic individuals have more hospital emergency room visits, and more severe respiratory symptoms, when particulate pollution levels increase.[30]

No investigations to date have measured the effects of particulates on submaximal or maximal exercise performance, and only one early study included particle exposure in the research design.[1, 6] In 1948, before human research review panels were prevalent at institutions, three healthy males inhaled aluminum dust (0.7 μm average particle diameter) three times, between brief periods of breathing clean air. The investigators concluded that bronchoconstriction and breathing frequency increased with each successive dose of aluminum dust.[31]

Aerosols

The four previous substances (CO, NO_2, SO_2, and fine particulates) are primary pollutants, which exert their influence on humans exactly as they are issued from the source. This section marks the transition to three categories of *secondary* pollutants, which are formed by interactions of primary pollutants with other compounds, ultraviolet light, or with each other. Aerosols have been classified both as particulates[2, 24] and as a distinct pollutant category.[6] They are considered separately in this section because they meet the criteria of a secondary pollutant; that is, they can be formed when sulfur oxides or nitrogen oxides react with photochemical products and airborne particles.

An **aerosol** is a suspension of ultramicroscopic solid or liquid particles in air or another gas; examples include smoke, fog, or mist. Thus, fine particulates (see previous section) may be inhaled in the form of an aerosol. Some aerosols are soluble and can enter the blood, resulting in effects not related to the pulmonary system of the body.[2]

Numerous aerosol studies have involved exercise. The majority of these have investigated sulfate, sulfuric acid (H_2SO_4), or nitrate aerosols. The sulfate aerosols arise from ammonia combining with H_2SO_4 to form ammonium sulfate ($[NH_4]_2SO_4$) and ammonium bisulfate ($NH_4H_2SO_4$). The nitrate aerosol most often studied is ammonium nitrate (NH_4NO_3).[6]

The sulfate aerosols ($(NH_4)_2SO_4$ and $NH_4H_2SO_4$ apparently have no or little effect on pulmonary function measure or clinical symptoms of exercising normal and asthmatic test subjects.[32, 33] Inhalation of H_2SO_4 aerosols during exercise, in contrast, appears to cause measurable changes in pulmonary function, despite the findings of early studies (i.e., before 1981) to the contrary.[2, 34, 35] Recently, Linn and colleagues exposed children to H_2SO_4 aerosols during exercise, in two studies. The first involved 41 subjects, aged 9 to 12 years, 26 of whom had allergy or mild asthma.[36] The acid dose (0.6 μm H_2SO_4 aerosol, 100 μg/m^3) did not affect the healthy children, but was positively correlated with clinical symptoms in the allergy/asthma subgroups. No changes were noted in spirometry measurements of lung function or perceived discomfort in either group. The second study exposed 24 asthmatic volunteers, aged 11 to 18 years, to an H_2SO_4 aerosol slightly more concentrated than the one described above (127 μg/m^3). Exercise-induced bronchospasm was observed at the end of three 15 min exercise sessions. Additionally, evidence suggested that a subgroup of these children were susceptible to acid pollution for unknown reasons.[37] Please note that neither study assessed maximal exercise performance.

Exposure to H_2SO_4 aerosol in high concentration for prolonged periods, multiple exposures, or with larger aerosol particles, may cause this pollutant to have adverse effects on humans.[6] A rat model of pulmonary toxicology has identified the nature of potential tissue and cellular injury caused by inhalation of this aerosol. Microscopic examination of damage to different cell types demonstrated death, followed by a wave of new cell replication in the surface cells (epithelium) of the nasal passage. When a mixture of this pollutant and ozone was inhaled, additional damage was observed in deep lung cells.[38]

With regard to nitrate aerosols, two studies have shown no substantial changes in pulmonary function tests or clinical symptoms in either healthy or asthmatic test subjects.[32, 39] The NH_4NO_3 concentrations were 80 and 200 $\mu g/m^3$, respectively. Exercise sessions involved 2 h of light, intermittent cycling, and a 4 h exposure that included two 15 min treadmill walks. Thus, experimental evidence does not support major alterations in respiratory responses associated with NH_4NO_3 aerosol inhalation.[6]

Ozone

Ozone is produced in nature by the action of ultraviolet radiation (UVR) on oxygen as this radiation enters earth's atmosphere from outer space. Ozone is essential to life on earth because it filters various types of UVR that destroy living organisms. It is found in the greatest concentrations in the ozonosphere, a layer of the upper atmosphere that lies between 9.5 and 48 km (6 and 30 mi) above the earth's surface. In this layer, molecular oxygen (O_2) absorbs UVR, and the imparted energy breaks the molecule into two oxygen atoms. These released atoms may combine with O_2 to form **ozone (O_3)** or may recombine to form O_2. Thus, O_3 is a secondary pollutant.

In the air covering large cities, however, O_3 is a pollutant that is primarily formed by the action of sunlight (i.e., ultraviolet rays) on automobile exhaust fumes.[2] Photochemical smog results from chemical reactions involving hydrocarbons (compounds that contain only hydrogen and carbon) and nitrogen oxides in the presence of sunlight. Ozone is one of the main constituents of smog. Differences in the intensity of solar radiation explain why O_3 levels peak near midday year-round (figure 6.4, page 212), and are four to five times higher in summer than in winter months.[7]

In humans, O_3 inhalation impairs pulmonary function, causes respiratory discomfort, and increases the number of reported clinical symptoms.[40-43] These responses are exacerbated during exercise because (a) the absolute amount of O_3 inhaled increases, (b) the uniformity of ventilation throughout all lung tissue increases, and (c) "nasal scrubbing" (i.e., absorbing gases during quiet breathing through the nose) is compromised.[44] In fact, pulmonary physiologists often calculate the effect of exercise (and other factors that increase ventilation) on the amount of O_3 inhaled,[43, 45] by using the **effective dose**

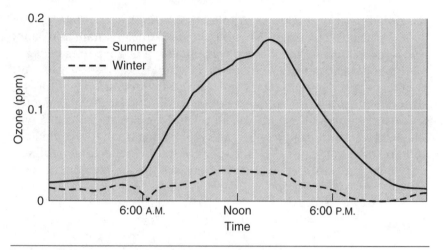

Figure 6.4 Daily fluctuations of ozone (O_3) during summer and winter in the Los Angeles area.
Reprinted from Pandolf 1988.

(ED) concept developed by Silverman and colleagues in 1976.[46] The ED is expressed as the multiplication product of O_3 concentration (ppm), $\dot{V}_E$ (liters per minute), and exposure time (minutes); ED is a unitless, relative term. These three factors interrelate to determine the total amount of O_3 reaching (i.e., the impact that O_3 has on) lung tissue. Interestingly, the O_3 concentration has been shown, by multiple regression statistical analysis, to be the most important of these three factors in predicting pulmonary function impairment.[41, 45, 47]

Federal air quality standards in the United States allow an O_3 exposure of 0.12 ppm over a 1 h period.[48] Some cities exceed these levels. Daily average maximums in the Los Angeles Basin, for example, have been as high as 0.23 ppm for 1 month, with a single-day maximum of 0.54-0.60 ppm O_3.[2, 49] Many cities announce an ozone alert when O_3 levels reach 0.20 ppm.[50] Previous investigations that have evaluated the effects of O_3 inhalation on exercise performance have involved levels that approximate and exceed these concentrations (i.e., 0.12-0.75 ppm O_3) for brief periods of time (i.e., 15-120 min),[51, 52] but the total ED is generally much lower in research studies than in real cities because test subjects are not exposed to high O_3 levels continuously, on consecutive days. Nevertheless, the research published to date reveals several important facts about the effects of ozone on exercise. These include the following 11 findings.

1. The most common clinical symptoms (in descending order of

occurrence) reported during three exercise studies were cough, shortness of breath, excessive sputum, throat tickle, raspy throat, nausea;[44] throat tickle, shortness of breath, headache, cough congestion;[43] and cough, shortness of breath, raspy throat, throat tickle, headache.[52] These studies involved air containing 0.12-0.30 ppm O_3 and represent ozone toxicity.[23]

2. Impaired PFT measurements (see "Tests of Pulmonary Function," page 200) occur during exercise at O_3 concentrations ranging from 0.12 to 0.24 ppm.[45, 47, 52-54] Maximal $\dot{V}_E$ during exercise may be a more sensitive indicator of the effects of O_3 exposure than standard PFT.[43]

3. Diminished endurance exercise performance has been observed at concentrations of 0.24 ppm[52] and 0.35 ppm O_3,[42, 53] but no change in exercise performance has been seen at concentrations of 0.15 and 0.30 ppm O_3[42] or 0.12 and 0.18 ppm O_3.[52] The ED and workloads in these experiments (duration: from 37 to 60 min) are difficult to compare, but the ozone levels suggest that a concentration threshold exists (e.g., 0.20-0.40 ppm O_3), beyond which endurance performance is negatively affected.

4. $\dot{V}O_2$max, the physiological variable most closely related to maximal endurance performance, decreased subsequent to one 50 min exercise bout while breathing 0.35 ppm O_3,[42] and subsequent to 2 h of exercise while breathing 0.75 ppm O_3.[51] In contrast, no change in $\dot{V}O_2$max occurred after breathing lower levels of ozone (0.15 and 0.30 ppm) for 37 min during an incremental exercise test.[43] As was noted in item 3 above for endurance performance, a concentration or ED threshold probably exists between 0.20 and 0.40 ppm O_3, beyond which $\dot{V}O_2$max decreases. Savin and Adams suggest that this threshold lies near 0.30 ppm O_3.[43] It also is likely that the effects of O_3 on $\dot{V}O_2$max and near-maximal exercise performance do not occur at low-exercise intensities.[6]

5. Subjective sensations of discomfort increase as the O_3 concentration of inhaled air increases from 0.0 to 0.3 ppm.[43] This supports the concept, proposed initially in 1985-1987,[43, 51, 55] that maximal work may be limited by psychological factors or clinical symptoms, rather than by physiological factors. Because most subjects can distinguish between control (i.e., pure filtered air) and treatment (i.e., O_3) tests,[43] it is likely that the enhanced perception of respiratory discomfort impairs maximal exercise performance.[42, 51, 56]

6. The ozone sensitivity of normal male college students is similar to that of endurance athletes during mild exercise.[52] This suggests that physical training per se does not alter one's response to O_3

exposure. However, because highly trained endurance athletes can sustain very high $\dot{V}_E$ for prolonged periods, they exhibit significant PFT impairment at low O_3 levels because their ED is very high.[52, 56]

7. The decrements in physical performance, due to increased subjective sensations and altered physiological responses, are accentuated when exercise is undertaken in a hot environment.[44] This apparently occurs for two reasons. First, $\dot{V}_E$ may be increased by ambient heat, which in turn may increase the ED. Second, a high ambient temperature may potentiate lung sensitivity to photochemical oxidants such as ozone via unknown mechanisms.

8. Comparisons of male and female test subjects indicate that women *may or may not* experience a substantially greater PFT impairment after O_3 inhalation than men, regardless of whether similar, relative (i.e., same percent $\dot{V}O_2$max) or absolute (i.e., different percent $\dot{V}O_2$max) workloads are employed in testing.[45, 54]

9. Distinct differences have been observed among healthy males who cycled for 1 h at 65% $\dot{V}O_2$max while breathing 0.30 ppm O_3. This great between-subject disparity in exercise performance hypothetically may be due to differences in the control of depth and frequency of breathing, or to differences in feelings of exertional breathlessness.[57]

10. Older men and women (51-76 years) were less responsive to O_3 inhalation than young men and women (18-26 years), based on PFT measurements after a 2 h ozone exposure (0.45 ppm) that included three 20 min exercise bouts.[58] Although speculative, it is possible that aging altered their O_3 receptors.

11. No racial differences in O_3 sensitivity are known, although the number of studies is very limited. One well-designed investigation compared black and white men and women, aged 18 to 35 years.[54] Over 90 subjects were in each of the four race-gender groups; they participated by inhaling pure filtered air (0.0 ppm O_3) and five different levels of ozone (0.12, 0.18, 0.24, 0.30, 0.40 ppm O_3) on six separate days. The O_3 exposure tests lasted for 2 h, including four 15 min treadmill walks. Although cough, FEV_1, and R_{aw} (see "Tests of Pulmonary Function," page 200) were altered by O_3 in all subjects, no differences were found between-race, in either men or women.

Anatomically and physiologically, the effects of ozone on the bronchial tubes and lungs apparently result from stimulation of irritant receptors (i.e., for O_3) in the pharynx region of the throat, behind the tongue, which in turn causes a hyperirritable state in the vagus nerve (see "Sulfur Dioxide," page 206) and involuntary inhibition of inspiration.[44, 59] This neurochemical response results in rapid, shallow breathing during O_3 and exercise exposure,[42, 44] as well as a reduction in the amount of ambient air that reaches the alveoli.[57] Both of these factors enhance one's sensation of breathlessness during exercise. The PFT measurements that indicate impairment (i.e., reduced capacity) are FVC, FEV_1, and R_{aw}.[59] Because these responses may alter the diffusion of O_2 from the alveoli into the capillary bloodstream, it is possible that this is one mechanism by which O_3 reduces exercise performance. Bates, in fact, proposed this mechanism (i.e., reduced alveolar ventilation-perfusion ratio) in 1980 along with two others: a decreased O_2 saturation in arterial blood, and an increased energy requirement of respiratory muscular effort.[60] However, subsequent tests of endurance athletes have shown that these three mechanisms were not the cause of the reduced exercise time during competitive laboratory simulations.[52] This does not eliminate the possibility, however, that these mechanisms work to undermine exercise performance in untrained healthy or asthmatic individuals.

Research Shows That Ozone Induces Inflammation

In the 1990s, one other response to O_3 exposure has been identified and acknowledged as important in the medical literature. This involves inflammation—the body's response to trauma, infection, or irritation.[4] At rest, O_3 exposure results in biochemical changes that stimulate the production of **neutrophil** cells and mediators of inflammation in the nasal cavity and upper airways, because as much as 40% of inspired ozone collects at those sites during quiet nose breathing.[24]

During exercise, inflammation also may occur deeper in the lung. A recent study attempted to clarify the unexplained relationship between O_3-induced airway inflammation and decreased inspiratory capacity.[59] Ten test subjects consumed either the nonsteroidal anti-inflammatory drug ibuprofen or a placebo pill, at 1.5 h before and at the midpoint of each experiment. The pollutant exposure (0.4 ppm O_3 for 2 h) included 1 h of intermittent exercise. Although ibuprofen had no effect on neutrophil production, it significantly reduced the levels of known mediators of inflammation (e.g., prostaglandin E_2, thromboxane B_2) and the levels of **interleukin-6**, a protein that enhances immmune responses. Because these changes were concurrent with a *blunting* of the usual decrease in FEV_1, this study supported a connection between the natural compounds that mediate inflammation and pulmonary impairment.

With regard to exposure levels, it has been shown that exercising exposure to 0.4 ppm O_3 for 2 h and exposure to 0.08 ppm O_3 for 6.6 h both resulted in biochemical changes similar to those described above. In the latter condition, increases in lung fluid neutrophils, prostaglandin E_2, fibronectin, interleukin-6, and a decrease in the phagocytic (engulfing) activity of alveolar macrophage cells all indicated obvious inflammatory and immune system reactions to ozone in the lung.[61] This 0.08 ppm O_3 concentration is considered to be a low-level, prolonged exposure that would be common in most large cities around the world.[48-50] It also suggests that virtually all previous exercise studies have induced some degree of inflammation in bronchial tubes or lung tissue, because these studies have utilized O_3 levels ranging from 0.12 to 0.75 ppm and have involved increased ventilation from 0.6 h to 2 h, indicating large effective doses.[42, 43, 45, 47, 51-53]

Could inflammatory responses be involved in some aspect(s) of the O_3-induced impairment of pulmonary function, maximal exercise performance, or $\dot{V}O_2$max? Yes, but clarification of the exact mechanism(s) awaits future research. It also is possible that edema (fluid accumulation, swelling), a hallmark of inflammation, may be involved. An early report suggested that some pulmonary edema occurred after a 2 h exposure to ozone in concentrations of 0.6-0.8 ppm.[62] Exercise-induced increases in blood pressure could further increase fluid movements into the space between cells (interstitial space) or into/between alveoli,[57] thereby potentially impairing oxygen and carbon dioxide gas diffusion across alveolar capillary membranes.

Adaptation to Ozone

Adaptive response to ozone (O_3) exposure, as indicated by several different lines of research, occurs in a way that is analogous to the adaptation described above for SO_2. This adaptive response was initially suggested by a comparison of residents of Los Angeles to residents of Montreal.[63] Since that time, the following research studies have suggested that adaptation (i.e., reduced pulmonary impairment and fewer clinical symptoms) occurs after successive O_3 exposures:[6]

• Six patients with airway hyperactivity were exposed at rest to ozone (0.5 ppm) for 2 h on 4 consecutive days. The pulmonary function decrements observed on days 1-3 were mostly reversed by day 4.[64]

• A series of experiments involving 75 test subjects (62 men, 13 women) employed intermittent exercise and a 2 h inhalation exposure (0.20-0.50 ppm O_3). As summarized by Dr. Lawrence Folinsbee, these experiments showed that adaptation to O_3 during exercise requires 2-5 days, remains for 7-20 days after exposures cease (depending on between-subject differences), and is similar in women and men.[65] Figure 6.5 depicts the number of days that were required to produce adaptation to 0.42 ppm O_3, during 2 h exposures, in 24 subjects.

• Adaptation to inhaled ozone may be lost completely after a 7-day interval without exposure.[66]

• Adaptation to ozone during exercise may be viewed as a potentially harmful response (because of damage to or loss of a normal defense mechanism).[6, 7, 23]

However, the human adaptive responses to ozone inhalation may not be as straightforward as figure 6.5 implies. An interesting study conducted at the University of California at Davis suggests that O_3 adaptation may be more complex than previously thought.[42] Eight trained males engaged in a 1 h cycling protocol while breathing 0.35 ppm O_3, which

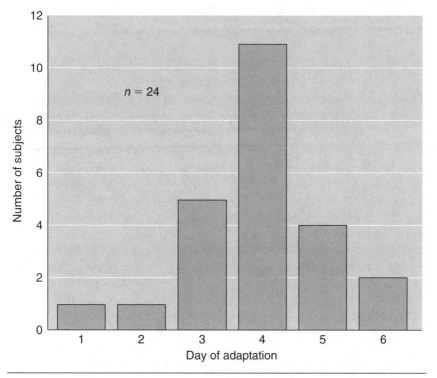

Figure 6.5 The number of days that 24 test subjects required to exhibit adaptation to 0.42 ppm O_3, during 2 h exposures on 5 consecutive days.
Reprinted from Folinsbee et al. 1983.

ended with a $\dot{V}O_2$max test on three occasions. The initial exposure to O_3 induced a marked increase in subjective symptoms of discomfort and illness (73 total reports of cough, shortness of breath, throat tickle, raspy throat, nausea, excessive sputum, etc.) when compared to pure, filtered air. Other effects included significant pulmonary function impairment (i.e., FVC, FEV_1, FEF) and decrements in maximal exercise performance time (from 253 to 211 sec), $\dot{V}_E$max (from 146 to 124 L/min), maximal heart rates (from 184 to 173 beats/min), and $\dot{V}O_2$max (from 3.85 to 3.62 L/min). This protocol was then repeated on four consecutive days; on the fourth daily exposure to O_3, exercise performance time (239 sec), $\dot{V}_E$ max (138 L/min), maximal heart rate (181 beats/min), $\dot{V}O_2$max (3.79 L/min), and subjective symptoms (24 total reports) were significantly improved (p < .05). However, all days of pulmonary function impairments remained similar to the first day, suggesting that no adaptation occurred in PFT measurements. Figure 6.6 illustrates these responses. Unfortunately, only the PFT measurements were taken on all days of O_3 exposure; also, measurements were taken for four consecutive days, which may not have been long enough because adaptation to O_3 exposure during exercise may require five days or more.[65]

Peroxyacetyl Nitrate

The third category of secondary pollutants is **peroxyacetyl nitrate (PAN)**, which, like ozone, is formed by a photochemical reaction and is one of the common constituents of smog.[67] It is believed that PAN serves as a stable storage form of NO_2 and other nitrogen oxides, allowing these highly reactive compounds to be transported in the colder, high regions of earth's atmosphere. Once transported, PAN can easily release free NO_2 in warmer, low-altitude air layers.[68] The quantities of PAN in the air of large cities depends on the intensity and duration of daily sunshine, air temperature, the concentration of PAN and other photochemical oxidants such as ozone at the beginning of the day, as well as the emission rates and concentrations of primary pollutants (e.g., NO_2). In fact, the increase in ozone in the late morning and early afternoon of hot summer days, as shown in figure 6.4 (page 212), is accompanied by an increase in PAN and a decrease in NO_2.[67]

PAN was initially identified in automobile exhaust as a pollutant in the early 1950s.[7] Since that time, it has been recognized as an eye irritant with a distinctive odor that results in blurred vision and eye fatigue at levels as low as 0.24 ppm.[69] In comparison, PAN concentrations in Los Angeles have exceeded 0.50 ppm when environmental conditions were favorable for a great buildup of photochemical smog.[2]

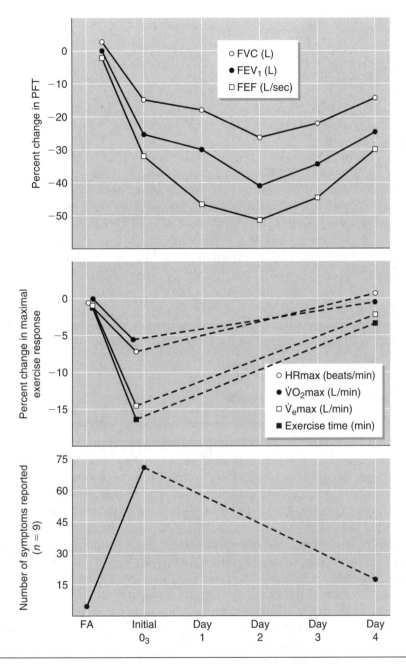

Figure 6.6 Changes in pulmonary function tests (PFT), maximal exercise responses, and symptoms reported. Days 1-4 represent consecutive daily O_3 exposures. FA, initial O_3, and day 1 were randomly presented, 3-4 days apart.

Reprinted from Folinsbee et al. 1983.

Only a few studies have investigated the effects of PAN during submaximal exercise.[6] One involved light treadmill walking (35% $\dot{V}O_2$max) lasting for 3.5 h, while breathing air containing 0.24 ppm PAN. The results showed no changes in oxygen consumption, ventilation, heart rate, rectal temperature, or skin temperature in both younger (22-26 years old) and older (45-55 years old) healthy males.[69] Values recorded during PFT (i.e., FEV_1) also revealed no change due to PAN exposure.[70]

Two studies have observed the effects of a PAN exposure of 0.27 ppm on $\dot{V}O_2$max.[12, 70] These experiments found no decrement in the $\dot{V}O_2$max values of two groups of smokers and nonsmokers: healthy younger (21-30 years) and older (40-57 years) men. Further, no changes in $\dot{V}_E$max, exercise time, maximal heart rate, or blood lactate concentration after exercise was observed. The investigators cautioned, however, that 0.27 ppm PAN might impair PFT or maximal exercise performance in unhealthy individuals (i.e., those with respiratory disease or compromised pulmonary function). In summary, the research regarding PAN exposure is sparse, particularly regarding maximal exercise performance. It appears unwise to form conclusions until higher concentrations are tested and persons with pulmonary impairment are evaluated.[6]

Mixtures of Air Pollutants

Because air is a mixture of gases, and because primary and secondary pollutants are produced continuously in large cities, it is difficult to distinguish the effects of individual pollutants outside of laboratories. Also, various pollutants may interact to pose a greater threat to health and exercise performance than that found with each single pollutant.[6] To date, interactions of pollutants have been categorized as either additive, synergistic, or antagonistic.[7] No antagonistic pollutant interactions have been observed.[6] In reevaluating the definitions of these categories, it seems appropriate to revise them slightly. Here, an *additive* interaction is one in which the total physiologic effect is the sum of the effects of two or more individual pollutants. A *synergistic* interaction involves a total physiologic effect that is greater than the sum of the individual pollutant effects.[6] A *benign* interaction is one in which no physiologic effects occur beyond those induced by one pollutant; this includes gas mixtures in which all components have no effect.

Table 6.2 summarizes numerous scientific and clinical research studies that have examined pollutant mixture exposures. Because the only

Table 6.2
Investigations of Air Pollutant Interactions

Year	Exercise intensity	Additive	Synergistic	Benign	References
1975, 1981, 1989	S			$O_3 + NO_2$	58, 73, 74
1975	S			$O_3 + NO_2 + CO$	73
1984	S			$O_3 +$ particulates	72
1983	S			$NO_2 +$ four aerosols	32
1983	S			$SO_2 +$ four aerosols	32
1983	S			$O_3 +$ cigarette smoke	71
1974, 1975	S, M			$CO + PAN$	12, 69
1975	S		$O_3 + SO_2$		76
1977, 1982	S			$O_3 + SO_2$	77, 78
1989, 1987	S			$O_3 + PAN$	58, 75
1989	S			$O_3 + PAN + NO_2$	58, 75
1987	S			$PAN + NO_2$	75

Abbreviations: S, submaximal; M, maximal; see text for chemicals involved.

Exposure to multiple pollutants, as may occur in an urban factory setting, may cause additive effects to your health.

study to observe either an additive or synergistic interaction[76] in healthy humans has been subsequently disputed,[77, 78] it is very doubtful that these categories actually exist, as presented in previous review articles.[6, 7] This absence of evidence implies that pollutant interactions of any type probably exist in theory only. An exception to this conclusion may be found in persons with asthma, as explained below.

MEDICAL CONCERNS: BRONCHIAL ASTHMA AND NASAL ALLERGIES

In addition to the previously mentioned complications, air pollution can exacerbate previously existing respiratory conditions. Among the most common are bronchial asthma and nasal allergies.

Bronchial Asthma

Bronchial asthma is an illness that involves hypersensitive airways and an inflammatory reaction that narrows the respiratory tree.[79] Its characteristic symptoms include shortness of breath, wheezing, and coughing. The increased resistance to airflow during asthma results from contraction of the smooth muscle surrounding the bronchial tubes, mucus secretion, and fluid seepage. Fine particulates (i.e., dust, cigarette smoke, pollen) may stimulate an asthma "attack." In some individuals, exercise stimulates bronchoconstriction (e.g., exercise-induced bronchoconstriction, exercise-induced asthma).

Nasal Allergy

A **nasal allergy** occurs when an airborne substance (e.g., pollen, dust mites) enters the nose and stimulates the production of antibodies and the release of the chemical histamine. Histamine causes inflammation, swelling, and increased fluid in the nose, leading to symptoms such as sinus headache, plugged ears, postnasal drip, and itchy eyes.[79] In contrast, the common cold often results in a fever, sore throat, thick nasal discharge, aches, and chills.

It is widely recognized that people who suffer from bronchial asthma, exercise-induced bronchoconstriction, or nasal allergies experience difficulties in air-polluted environments. This should not be surprising because the negative effects that NO_2, SO_2, O_3, fine particulates, and aerosols (see above) have on the respiratory system are similar or identical to the symptoms of bronchial asthma and nasal allergy. Table 6.3 presents the conclusions of ten research publications on the effects of air pollutants on persons with asthma. This table is summarized by the following statement: asthmatics experience a greater number (and severity) of symptoms of respiratory discomfort at lower pollutant concentrations than nonasthmatics.

The previous section of this chapter ended with the statement that no evidence verifies the existence of additive or synergistic pollutant interactions in healthy humans (table 6.2). However it has been shown, without subsequent evidence to the contrary, that at least one pollutant interaction may affect asthmatics. Eight male and five female adolescents performed moderate, intermittent exercise while inhaling

Table 6.3
Research Findings Regarding the Effects of Air Pollutants on Asthmatics

Pollutant	Nature of the effect	References
NO_2	A are more susceptible to NO_2 effects than NA.	6
SO_2	Ambient SO_2 level is related to shortness of breath in A; A experience bronchospasm after inhaling 0.2-0.5 ppm SO_2, whereas NA require >2 ppm SO_2.	6, 22, 80
Fine particulates	Ambient particulate level is related to coughing frequency and the incidence of clinical symptoms in A.	30, 80
O_3	A have greater bronchial inflammation than in NA; adaptation (e.g., tolerance) to repeated exposures is possible in A, although it is slower and less effective than in NA.	81, 82
H_2SO_4 aerosol	Ambient acid aerosol level is related to coughing frequency; an acid-pollution-susceptible subgroup exists among youthful A.	36, 37
PAN	Possible impairment of maximal exercise performance in A; both A and NA experience more severe respiratory symptoms.	7, 70

Abbreviations: NA, nonasthmatics; A, asthmatics.

0.12 ppm O_3 for 45 min, followed by a 15 min period of inhaling 0.1 ppm SO_2. This O_3 exposure significantly increased bronchial hyperresponsiveness to SO_2 (i.e., FEV_1 and other PFT measurements) at a level that ordinarily would not cause pulmonary impairment. This apparent pollutant interaction, and others, are worthy of further study.

Another promising line of research involves pharmacologic intervention before exposure to air pollutants and exercise. To date, three investigations have evaluated nonallergic asthmatics who inhaled sulfur dioxide during exercise.[83-85] The first used four 10 min exposures (0.75 ppm SO_2) at 1, 12, 18, and 24 h after application of the medication. The second investigation used SO_2 levels of 0.0, 0.5, and 1.0 ppm during 20 min of rest followed by 10 min of submaximal exercise. The third involved 10 min of heavy, continuous exercise and concentrations of 0.0, 0.3, and 0.6 ppm SO_2. The results of all studies showed, as expected, that exercise-induced bronchospasm increased as the SO_2 dose increased. Each of these investigations used an additional test, which exposed subjects to aerosol drug pretreatments, applied via a metered-dose inhaler, prior to exercise. The

© Terry Wild

Asthmatics exhibit a greater sensitivity to air pollution than nonasthmatics.

medications were salmeterol xinafoate,[83] ipratropium bromide,[84] and metaproterenol sulfate,[85] respectively. Salmeterol imparted clinically and statistically significant protection against bronchoconstriction for at least 12 h, and maintained an improvement in lung function for 18 h. Ipratropium bromide caused significant bronchodilation but did not completely protect these patients from the effects of SO_2 inhalation. Metaproterenol sulfate prevented bronchoconstriction during exercise at the 0.0 and 0.3 ppm levels, and greatly reduced this response at the 0.6 ppm level. Taken together, these three studies indicate that asthmatics can reduce respiratory impairment by using bronchodilating aerosol medications prior to exercise.

Epidemiological studies involving thousands of people have documented increasing evidences of asthma in various cities; these findings are believed to be related to O_3 levels. For example, hospital admissions for asthma correlate statistically with ambient O_3 concentrations.[50] It is unfortunate that recent evidence suggests that federal air quality guidelines (0.12 ppm O_3, see table 6.4) may exceed the threshold at which inflammation occurs deep within the lung[59] and in bronchial tubes,[50] where asthma and chronic bronchitis originate. Most cities wait to announce a smog alert until O_3 levels exceed 0.20 ppm. This concern is held for other pollutants as well. Contemporary levels of SO_2 and sulfuric acid aerosols result in adverse physiologic and clinical responses, in healthy individuals and in pollutant-sensitive groups such as asthmatics, children, and adolescents. Further, exercise may potentiate pollutant effects on symptoms, lung function, mucociliary clearance, and inflammation.[86]

Nasal allergies can be controlled to some extent through pharmacologic means. Antihistamine drugs (e.g., hay fever medications), for example, can block the nasal symptoms of SO_2 inhalation effectively.

Table 6.4
Air Quality Standards of the United States [48]

Pollutant	Allowable standard
CO	35 ppm
NO_2	0.05 ppm
SO_2	0.14 ppm
Fine particulates	150 $\mu g/m^3$
O_3	0.12 ppm

Also, therapy to optimize natural immune function may decrease the intensity of an allergic response to pollutants.[24] Thus, individuals with allergies are wise to seek help from a sports medicine physician who is experienced in the care of exercise-induced bronchospasm (EIB). EIB is the major problem of allergic athletes because the broncho-spasm often is under-appreciated or misinterpreted; it occurs in over 90% of asthmatics. For optimal exercise performance, EIB requires specific medical management. Although beyond the scope of this book, thorough descriptions of protocols to identify and treat EIB have been published elsewhere.[23]

Other types of allergic problems may degrade exercise performance. These include rhinitis, hay fever, sinusitis, hives, urticaria, and ad-verse reactions to drugs.[87] The management of these specific illnesses involves identifying them and providing supportive therapy, although measures have limited effectiveness in reducing the severity and du-ration of symptoms.

Recommendations for Counteracting Air Pollutants

As the paragraphs above have described, each component gas in pol-luted air should be viewed as a distinct stressor that acts on the air-ways and lungs uniquely. However, when exercising in a large city, you do not have the luxury of preparing for each pollutant separately; nor would that be practical. You inhale air as a mixture of many gases and pollutants, which cannot be filtered selectively. Therefore, the wisest approach to minimizing the impact of pollutants on respiratory func-tion and physical performance is to observe the following guidelines. These ideas will help you to limit your intake of all pollutants.[88-90]

- Listen to weather and news reports for daily updates of local air quality. If you are aware of thermal inversion layers and other meteorological conditions that increase the risk of high pollut-ant levels, you can take action when it is needed.

- When ambient O_3 levels are likely to be elevated, run in the morn-ing or at night. Remember that O_3 levels peak at approximately 3 P.M. each day (figure 6.4, page 212). Unfortunately, CO levels are lowest at the time that O_3 levels are highest, and peak twice each day (figure 6.7), at approximately 7 A.M. and 8 P.M. This paradox may require that you determine whether O_3 or CO will impair your performance the most, and act accordingly. Also, remember that CO peaks because of automobile exhaust production during the morning and afternoon rush hours, and that O_3 peaks when the sun's ultraviolet radiation is most intense, favoring the photochemical reactions that produce smog.

- If high O_3 or SO_2 levels are anticipated, you should exercise for 4 to 6 days under the same exercise and environmental conditions that you anticipate during an upcoming competition or other lengthy exposure. Because the lungs and bronchial tubes adapt to these two pollutants, your respiratory system will become desensitized to O_3 or SO_2 exposure. This effect may last for only a few days, however, so plan carefully. If air quality improves for several days, then worsens on the target date, you may have to lower your performance expectations.

- Select training courses in parks or along the ocean, if possible, where breezes whisk away automobile and industrial exhausts laden with CO and other pollutants.

- Minimize your exposure to pollutants while en route to an event. Consider closing windows and limiting the air intake of your vehicle.

- Minimize the warm-up period, to limit your exposure to polluted air.

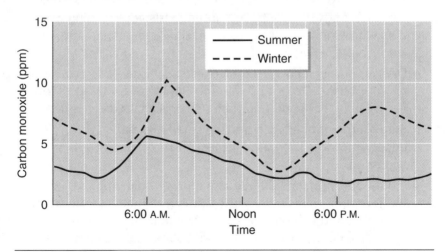

Figure 6.7 Daily fluctuations of carbon monoxide (CO) during summer and winter in the Los Angeles area.
Reprinted from Pandolf 1988.

- Avoid congested highways and intersections during training, stay at least 10-15 m away from the exhaust pipes of cars and trucks, and consider wind direction so that you ride or run on the up-wind side of the road. There is no need to be obsessive about pollution; it is a fact of life in cities. Simply use common sense.

- Consider the effective dose (ED) of training sessions. You will re-call (see section titled "Ozone," page 211) that ED is a relative term, expressed without units, and calculated as the multiplication product of O_3 concentration (ppm), $\dot{V}_E$ (liters per minute), and exposure time (minutes); the unit of ED is ppm per liter O_3. Frykman provides an interesting perspective on ED, by compar-ing an 8 h sedentary exposure to one 30 min intense workout.[3] His concept appears below:

$$\text{Sedentary:} (10) \cdot (6) \cdot (480) = 28{,}800$$

$$\text{Exercise:} (10) \cdot (100) \cdot (30) = 30{,}000$$

These calculations illustrate that your ED during one 30 min train-ing session can exceed the dose that you receive during 8 h of sedentary living. It is obvious that exercise intensity usually will be the most important factor in determining your ED of a pollut-ant. Thus, reducing the intensity of workouts (e.g., reducing run-ning speed) will lower $\dot{V}_E$ and the ED effectively. This can be accomplished by incorporating low-intensity technique/skill train-ing on smoggy days, or by performing resistance training work-outs with high loads (i.e., nearly maximal), few repetitions, and lengthy rest periods.

- Daily air-quality reports rate the effects that local pollutants have on your health. Do not exercise on days that are rated as "un-healthful," or "hazardous," or when a "smog alert" has been issued."[89]

- Stop exercise immediately and seek medical advice if you experi-ence symptoms such as tightness in the chest, coughing, or wheezing.[89]

- Indoor environments often contain lower pollutant levels than outdoor sites. This is especially helpful to asthmatics.[24] However, in rare situations, high levels of certain pollutants can be found indoors. A classic example of this is a case in which a group of young ice hockey players experienced CO poisoning at an indoor rink. The source of the pollution was an ice resurfacing machine that was powered by gasoline.[1,6,91]

- Sidestream smoke from the tip of a cigarette contains more CO than smoke that is directly inhaled. Breathing sidestream smoke or smoking cigarettes immediately after heavy exercise can accelerate the movement of CO into the bloodstream.[92] You should avoid secondhand cigarette smoke before and after exercise.

- Asthmatics should observe a few nonpharmacologic precautions to minimize the duration and intensity of bronchospasm during exercise.[24, 92] First, they should exercise in a pool or other warm-humid environments. Second, they should exercise at a low intensity. Third, they should wear a scarf or disposable respiratory mask in cold environments to warm the air and limit pollutant intake; nose breathing (versus mouth breathing) accomplishes these same goals. The cold-air sensitivity of asthmatics is widely recognized.[92] Finally, workouts should begin with 5-10 min of light stretching and breathing exercises, continue with a gradual buildup of intensity (e.g. 5-10 min), and end with a 10-20 min cool-down.

REFERENCES

1. Frykman, P.N. 1988. Effects of air pollution on human exercise performance. *Journal of Applied Sport Science Research 2(4)*: 66-71.

2. Horvath, S.M. 1981. Impact of air quality in exercise performance. *Exercise and Sport Sciences Reviews 9*: 265-296.

3. Shaheen, E. 1974. *Environmental pollution awareness and control.* Mahomet, IL: Technology Incorporated, 27-44.

4. Haagenson, D. 1979. Meteorological and climatic factors affecting Denver air quality. *Atmospheric Environment 13*: 79-85.

5. Guyton, A.C., & Hall, J.E. 1996. *Textbook of medical physiology.* 9th ed. Philadelphia: Saunders.

6. Pandolf, K.B. 1988. Air quality and human performance. In *Human performance physiology and environmental medicine at terrestrial extremes*, edited by K.B. Pandolf, M.N. Sawka, & R.R. Gonzalez. Indianapolis: Benchmark Press.

7. McCafferty, W.B. 1981. *Air pollution and athletic performance.* Springfield, IL: Charles C Thomas.

8. Chihaia, V.; Jonsecu, C.; & Konsecu, A. 1974. Etiological aspect of occupational intoxication by carbon monoxide. *Igiena 23*: 49-53.

9. Chiodi, H.; Dill, D.B.; Consolazio, F.; & Horvath, S.M. 1941. Respiratory and circulatory response to acute carbon monoxide poisoning. *American Journal of Physiology 134*: 683-693.

10. Aronow, W.S. 1978. Effects of passive smoking on angina pectoris. *New England Journal of Medicine 299*: 21-24.

11. Horvath, S.M.; Raven, P.B.; Dahms, T.E.; & Gray, D.J. 1975. Maximal aerobic capacity at different levels of carboxyhemoglobin. *Journal of Applied Physiology 38*: 300-303.

12. Raven, P.B.; Drinkwater, B.L.; Ruhling, R.O.; Bolduan, N.; Taguchi, S.; Gliner, J.; & Horvath, S.M. 1974. Effects of carbon monoxide and peroxyacetyl nitrate on man's maximal aerobic capacity. *Journal of Applied Physiology 36*: 288-293.

13. Laties, V.G., & Merigan, W.H. 1979. Behavioral effects of carbon monoxide on animals and man. *Annual Review of Pharmacology and Toxicology 19*: 357-392.

14. Nicholson, J.P., & Case, D.B. 1983. Carboxyhemoglobin levels in New York City runners. *The Physician and Sportsmedicine 11(3)*: 134-138.

15. Horvath, E.P.; do Pico, G.A.; Barbee, R.A.; & Dickie, H.A. 1978. Nitrogen dioxide induced pulmonary disease. *Journal of Occupational Medicine 20*: 103-110.

16. Kim, S.U.; Koenig, J.Q.; Pierson, W.E.; & Hanley, Q.S. 1991. Acute pulmonary effects of nitrogen dioxide exposure during exercise in competitive athletes. *Chest 99*: 815-819.

17. Mustafa, M.G., & Tierney, D.F. 1978. Biochemical and metabolic changes in the lung with oxygen, ozone, and nitrogen dioxide toxicity. *American Review of Respiratory Disease 118*: 1061-1090.

18. Halleday, R.; Huberman, D.; Blomberg, A.; Stjernberg, N.; & Sandstrom, T. 1995. Nitrogen dioxide exposure impairs the frequency of the mucociliary activity in healthy subjects. *European Respiratory Journal 8*: 1664-1668.

19. Nadel, J.A.; Salem, H.; Tample M.; & Tokewa, Y. 1965. Mechanism of bronchoconstriction during inhalation of sulfur dioxide. *Journal of Applied Physiology 20*: 164-171.

20. Frank, N.R.; Amdur, M.O.; Worcester, J.; & Whittenberger, J.L. 1962. Effects of acute controlled exposure to SO_2 on respiratory mechanics in healthy male adults. *Journal of Applied Physiology 17*: 252-258.

21. Speizer, F.E., & Frank, R. 1966. The uptake and release of SO_2 by the human nose. *Archives of Environmental Health 12*: 725-728.

22. Kleinman, M.T. 1984. Sulfur dioxide and exercise: Relationships between response and absorption in upper airways. *Journal of Air Pollution Control Association 34*: 32-37.

23. Pierson, W.E.; Covert, D.S.; Koenig, J.Q.; Namekata, T.; & Kim, Y.S. 1986. Implications of air pollution effects on athletic performance. *Medicine & Science in Sports & Exercise 18*: 322-327.

24. Gong, H., & Krishnareddy, S. 1995. How pollution and airborne allergens affect exercise. *The Physician and Sportsmedicine 23*: 35-43.

25. Amdur, M.O.; Melvin, W.W.; & Drinker, P. 1953. Effects of inhalation of sulfur dioxide in man. *Lancet 1*: 758-759.

26. Anderson, I.; Lundqvist, G.R.; Jensen, P.L.; & Proctor, D.F. 1974. Human response to controlled levels of sulfur dioxide. *Archives of Environmental Health 28*: 31-39.

27. Rabinovitch, S.; Greyson, N.D.; Weiser, W.; & Hoffstein, V. 1989. Clinical and laboratory features of acute sulfur dioxide inhalation poisoning. *American Review of Respiratory Disease 139*: 556-558.

28. Corn, M., & Burton, G. 1967. The irritant potential of pollutants in the atmosphere. *Archives of Environmental Health 14*: 54-61.

29. Kilburn, K. 1967. Cilia and mucous transport as determinants of the response of lung to air pollutants. *Archives of Environmental Health 14*: 77-91.

30. Dockery, D.W., & Pope, C.A. 1994. Acute respiratory effects of particulate air pollution. *Annual Review of Public Health 15*: 107-132.

31. Dantrebande, L.; Alford, W.C.; Highman, B.; Downing, R.; & Weaver, F.L. 1948. Studies on aerosols v: Effect of dust and pneumodilating aerosols on lung volume and type of respiration in man. *Journal of Applied Physiology 1*: 339-349.

32. Stacy, R.W.; Seal, E.; House, D.E.; Green, J.; Roger, L.J.; & Raggio. 1983. A survey of effects of gaseous and aerosol pollutants on pulmonary function of normal males. *Archives of Environmental Health 38*: 104-115.

33. Avol, E.L.; Jones, M.P.; Bailey, R.M.; Chang, N.N.; Kleinman, M.T.; Linn, W.S.; Bell, K.A.; & Hackney, J.D. 1979. Controlled exposures of human volunteers to sulfate aerosols. *American Review of Respiratory Disease 120*: 319-327.

34. Sackner, M.A.; Cipley, J.; Perez, D.; Ford, D.; Kwoka, M.; Reinhart, M.; Fernandez, R.; Michaelson, E.D.; Schreck, R.; & Wanner, A. 1978. Effect of sulfuric acid aerosol on cardiopulmonary function of dogs, sheep, and humans. *American Review of Respiratory Disease 118*: 497-510.

35. Horvath, S.M.; Folinsbee, L.J.; & Bedi, J. 1980. Effects of sulfuric acid mist exposure on pulmonary function. *Report to the Environmental Protection Agency.* University of California, Berkeley, CA.

36. Linn, W.S.; Gong, H.; Shamoo, D.A.; Anderson, K.R.; & Avol, E.L. 1997. Chamber exposures of children to mixed ozone, sulfur dioxide, and sulfuric acid. *Archives of Environmental Health 52*: 179-187.

37. Linn, W.S.; Anderson, K.R.; Shamoo, D.A.; Edwards, S.A.; Webb, T.L.; Hackney, J.D.; & Gong, H. 1993. Controlled exposures of young asthmatics to mixed oxidant gases and acid aerosol. *American Journal of Respiratory and Critical Care Medicine 152*: 885-891.

38. Kleinman, M.T.; Phalen, R.F.; Mantz, W.J.; Mannix, R.C.; McClure, T.R.; & Crocker, T.T. 1989. Health effects of acid aerosols formed by atmospheric mixtures. *Environmental Health Perspectives 79*: 137-145.

39. Kleinman, M.T.; Linn, W.S.; Bailey, R.M.; Jones, M.P.; & Hackney, J.D. 1980. Effect of ammonium nitrate aerosol on human respiratory function and symptoms. *Environmental Research 21*: 317-326.

40. Bates, D.V.; Bell, G.M.; Burnham, C.D.; Hazucha, M.; Mantha, J.; Pengelly, L.D.; & Silverman, F. 1972. Short-term effects of ozone on the lungs. *Journal of Applied Physiology 32*: 176-181.

41. Adams, W.C.; Savin, W.M.; & Christo, A.E. 1981. Detection of ozone toxicity during continuous exercise via the effective dose concept. *Journal of Applied Physiology: Respiratory, Environmental, and Exercise Physiology 51*: 415-422.

42. Foxcroft, W.J., & Adams, W.C. 1986. Effects of ozone exposure on four consecutive days on work performance and $\dot{V}O_2$max. *Journal of Applied Physiology 61*: 960-966.

43. Savin, W.M., & Adams, W.C. 1979. Effects of ozone inhalation on work performance and $\dot{V}O_2$max. *Journal of Applied Physiology 46*: 309-314.

44. Gibbons, S.I., & Adams, W.C. 1984. Combined effects of ozone exposure and ambient heat on exercising females. *Journal of Applied Physiology: Respiratory, Environmental, and Exercise Physiology 57*: 450-456.

45. Lauritzen, S.K., & Adams, W.C. 1985. Ozone inhalation effects consequent to continuous exercise in females: Comparison to males. *Journal of Applied Physiology 59*: 1601-1606.

46. Silverman, F.; Folinsbee, L.; Barnard, J.; & Shephard, R.J. 1976. Pulmonary function changes in ozone interactions of concentration and ventilation. *Journal of Applied Physiology 41*: 859-864.

47. Folinsbee, L.J.; Drinkwater, B.L.; Bedi, J.F.; & Horvath, S.M. 1978. The influence of exercise on the pulmonary function changes due to low concentrations of ozone. In *Environmental stress: Individual human adaptations*, edited by L.J. Folinsbee, J.A. Wagner, J.F. Borgea, B.L. Drinkwater, J.A. Gliner, & J.F. Bedi, 125-145. New York: Academic Press.

48. United States Government. 1984. *Code of federal regulations*. Title 40, Part 50. Washington, DC: U.S. Government Printing Office.

49. Mosher, J.C.; MacBeth, W.G.; Leonard, M.J.; Mullins, T.P.; & Brunelle, M.F. 1970. The distribution of contaminants in the Los Angeles Basin resulting from atmospheric reactions and transport. *Journal of the Air Pollution Control Association 20*: 35-42.

50. Adler, T. 1993. Health effects of smog: Worse than thought. *Science News 144*: 326.

51. Folinsbee, L.J.; Silverman, F.; & Shephard, R.J. 1975. Exercise responses following ozone exposure. *Journal of Applied Physiology 42*: 531-536.

52. Schelegle, E.S., & Adams, W.C. 1986. Reduced exercise time in competitive simulations consequent to low level ozone exposure. *Medicine & Science in Sports & Exercise 18*: 408-414.

53. Adams, W.C., & Schelegle, E.S. 1983. Ozone and high ventilation effects on pulmonary function and endurance performance. *Journal of Applied Physiology 55*: 805-812.

54. Seal, E.; McDonnell, W.F.; House, D.E.; Salaam, S.A.; Dewitt, P.J.; Butler, S.O.; Green, J.; & Raggio, L. 1993. The pulmonary response of white and black adults to six concentrations of ozone. *American Review of Respiratory Disease 147*: 804-810.

55. Folinsbee, L.J.; Silverman, F.; & Shephard, R.J. 1977. Decrease of maximum

work performance following O_3 exposure. *Journal of Applied Physiology: Respiratory, Environmental, and Exercise Physiology 42*: 531-536.

56. Adams, W.C. 1987. Effects of ozone exposure at ambient air pollution episode levels on exercise performance. *Sports Medicine (New Zealand) 4*: 395-424.

57. DeLucia, A.J., & Adams, W.C. 1977. Effects of O_3 inhalation during exercise on pulmonary function and blood biochemistry. *Journal of Applied Physiology: Respiratory, Environmental, and Exercise Physiology 43*: 75-81.

58. Dreschler-Parks, D.M.; Bedi, J.F.; & Horvath, S.M. 1989. Pulmonary function responses of young and older adults to mixtures of O_3, NO_2, and PAN. *Toxicology and Industrial Health 5*: 505-517.

59. Hazucha, M.J.; Madden, M.; Pape, B.; Becker, S.; Devlin, R.B.; Koren, H.S.; Kehrl, H.; & Bromberg, P.A. 1996. Effects of cyclo-oxygenase inhibition of ozone-induced respiratory inflammation and lung function change. *European Journal of Applied Physiology 73*: 17-27.

60. Bates, D.V. 1980. Effects of irritant gases on maximal exercise performance. In *Exercise bioenergetics and gas exchange*, edited by P. Cerretelli & B.J. Whipp, 337-344. Amsterdam, Netherlands: Elsevier/North-Holland Biomedical Press.

61. Devlin, R.B.; McDonnell, W.F.; Mann, R.; Becker, S.; House, D.E.; Schreinemachers, D.; & Koren, H.S. 1991. Exposure of humans to ambient levels of ozone for 6.6 hours causes cellular and biochemical changes in the lung. *American Journal of Respiratory Cell and Molecular Biology 4*: 72-81.

62. Young, W.A.; Shaw, D.B.; & Bates, D.V. 1967. Effects of low concentrations of ozone on pulmonary functions in man. *Journal of Applied Physiology 19*: 765-768.

63. Hackney, J.D.; Linn, W.S.; Karuza, S.K.; Buckley, R.D.; Law, D.C.; Bates, D.V.; Hazucha, M.; Pengelly, L.D.; & Silverman, F. 1977. Effects of ozone exposure in Canadians and Southern Californians: Evidence for adaptation? *Archives of Environmental Health 32*: 110-116.

64. Hackney, J.D.; Linn, W.S.; Mohler, J.G.; & Collier, C.R. 1977. Adaptation to short-term respiratory effects on ozone in men exposed repeatedly. *Journal of Applied Physiology 43*: 82-85.

65. Folinsbee, L.J.; Bedi, J.F.; Gliner, J.A.; & Horvath, S.M. 1983. Concentration dependence of pulmonary function adaptation to ozone. In *The Biomedical effects of ozone and related photochemical oxidants*, edited by M.A. Mehlman, S.D. Lee, & M.G. Mustafa, 175-187. Princeton Junction, NJ: Princeton Scientific.

66. Linn, W.S.; Medway, D.A.; Anzar, U.T.; Valencia, L.M.; Spier, C.E.; Tsao, F.S.D.; Fischer, D.A.; & Hackney, J.D. 1982. Persistence of adaptations to ozone in volunteers exposed repeatedly for six weeks. *American Review of Respiratory Disease 125*: 491-495.

67. Moser, W. 1986. Formation and transport of ozone and other photochemical oxidants. *Soz-Praventivmed (German) 31*: 48-52.

68. Singh, H.B.; Salas, L.J.; & Viezce, W. 1986. Global distribution of peroxyacetyl nitrate. *Nature 321*: 588-591.

69. Gliner, J.A.; Raven, P.B.; Horvath, S.M.; Drinkwater, B.L.; & Sutton, J.C. 1975. Man's physiological response to long-term work during thermal and pollutant stress. *Journal of Applied Physiology 39*: 628-632.

70. Raven, P.B.; Gliner, J.A.; & Sutton, J.C. 1976. Dynamic lung function changes following long-term work in polluted environments. *Environmental Research 12*: 18-25.

71. Shephard, R.J.; Urch, B.; Silverman, F.; & Corey, P.N.J. 1983. Interaction of ozone and cigarette smoke exposure. *Environmental Research 31*: 125-137.

72. Avol, E.L.; Linn, W.S.; Venet, T.G.; Shamoo, D.A.; & Hackney, J.D. 1984. Comparative respiratory effects of ozone and ambient oxidant pollution exposure during heavy exercise. *Journal of the Air Pollution Control Association 34*: 804-809.

73. Hackney, J.D.; Linn, W.S.; Law, D.C.; Karuza, S.K.; Greenberg, H.; Buckley, R.D.; & Pedersen, E.E. 1975. Experimental studies on human health effects of air pollutants III: Two-hour exposure to ozone alone and in combination with other pollutant gases. *Archives of Environmental Health 30*: 385-390.

74. Folinsbee, L.J.; Bedi, J.F.; & Horvath, S.M. 1981. Combined effects of ozone and nitrogen dioxide on respiratory function in man. *American Industrial Hygiene Association Journal 42*: 534-541.

75. Drechsler-Parks, D.M. 1987. Effect of nitrogen dioxide, ozone, and peroxyacetyl nitrate on metabolic and pulmonary function. *Research Reports of the Health Effects Institute 6*: 1-37.

76. Hazucha, M., & Bates, D.V. 1975. Combined effect of ozone and sulfur dioxide on human pulmonary function. *Nature 257*: 50-51.

77. Bell, K.A.; Linn, W.S.; Hazucha, M.; Hackney, J.D.; & Bates, D.V. 1977. Respiratory effects of exposure to ozone plus sulfur dioxide in Southern Californians and Eastern Canadians. *American Industrial Hygiene Association Journal 38*: 696-706.

78. Bedi, J.F.; Horvath, S.M.; & Folinsbee, L.J. 1982. Human exposure to sulfur dioxide and ozone in a high temperature-humidity environment. *American Industrial Hygiene Association Journal 43*: 26-30.

79. Armstrong, L.E. 1990. How are exercise and the environment related to nasal allergy and asthma? *National Strength and Conditioning Association Journal 12*: 85-86.

80. Ostro, B.D.; Lipsett, M.J.; Wiener, M.B.; & Selner, J.C. 1991. Asthmatic responses to airborne acid aerosols. *American Journal of Public Health 81*: 694-702.

81. Scannell, C.; Chen, L.; Aris, R.M.; Tager, I.; Christian, D.; Ferrando, R.; Welch, B.; Kelly, T.; & Balmes, J.R. 1996. Greater ozone-induced inflammatory responses in subjects with asthma. *American Journal of Respiratory Critical Care Medicine 154*: 24-29.

82. Gong, H.; McManus, M.S.; & Linn, W.S. 1997. Attenuated response to repeated

daily ozone exposures in asthmatic subjects. *Archives of Environmental Health 52*: 34-41.

83. Gong, H.; Linn, W.S.; Shamoo, D.A.; Anderson, K.R.; Nugent, C.A.; Clark, K.W.; & Lin, A.E. 1996. Effect of inhaled salmeterol on sulfur dioxide-induced bronchoconstriction in asthmatic subjects. *Chest 110*: 1229-1235.

84. McManus, M.S.; Koenig, J.Q.; Altman, L.C.; & Pierson, W.E. 1989. Pulmonary effects of sulfur dioxide exposure and ipratropium bromide pretreatment in adults with nonallergic asthma. *Journal of Allergy and Clinical Immunology 83*: 619-626.

85. Linn, W.S.; Avol, E.L.; Shamoo, D.A.; Peng, R.C.; Smith, M.N.; & Hackney, J.D. 1988. Effect of metaproterenol sulfate on mild asthmatics' response to sulfur dioxide exposure and exercise. *Archives of Environmental Health 43*: 399-406.

86. Gong, H. 1992. Health effects of air pollution. A review of clinical studies. *Clinical Chest Medicine 13*: 201-214.

87. Katz, R.M. 1984. Rhinitis and the athlete. *Journal of Allergy and Clinical Immunology 73*: 708-712.

88. American Lung Association of Massachusetts. 1989. *Air quality index for Massachusetts*. Boston: American Lung Association.

89. American Lung Association. 1989. *Air pollution tips for exercisers*. Washington, DC: American Lung Association.

90. Kukula, K. 1988. Pollution solutions. *Runners' World 23(8)*: 16.

91. American Medical Association. 1984. Carbon monoxide intoxication associated with use of a resurfacing machine at an ice-skating rink. *Journal of the American Medical Association 251*: 1016.

92. Armstrong, L.E. 1991. The role of cold air in the onset of exercise-induced bronchospasm. *National Strength and Conditioning Association Journal 13(5)*: 1991.

chapter 7

Weather Patterns and Air Ions

The "Witches' Winds" come sweeping down. Called the Santa Ana in southern California, the Chinook in Canada, the Foehn in central Europe, the Sharav in Israel, and the Mistral in France, the winds are said to come on the heels of an increase in illness, domestic quarrels, suicides, murders and accidents. One theory blames this phenomenon on the positive ions that are generated by, but move ahead of, the hot winds.

The plunging, white-trimmed streams of the waterfall are mesmerizing, soothing. Some would attribute this effect to visual and auditory beauty; others would add a third factor—the ability of the falling water to spew negative ions into the surrounding air.

—Linda Garmon, 1981[1]

Each week, you experience changes in atmospheric factors as weather fronts pass through your community. It is reasonable, given the many factors that alter mood state, to expect that changing weather patterns might alter internal homeostasis and affect health and physical or mental performance. Although major fluctuations in temperature, humidity, wind speed, solar radiation, and air pressure were considered in previous chapters, this chapter focuses on air ionization—a specific environmental factor that has been studied for over 40 years. Although the experimental designs and results of air ion studies must be critically examined, and although research is not conclusive, it is possible that subtle effects on health and performance exist. These effects apparently are greater for "sensitive" individuals during major changes in local weather conditions.

BIOMETEOROLOGY

Chapter 1 explained the nature of cell homeostasis and the means by which the human body restores balance after external stressors disrupt the internal environment. The present chapter describes ways that weather and climate affect health and human performance. It is important because (a) we all experience changes in weather and (b) all strong environmental stressors have some effect on the immune system via the CNS, and most involve the endocrine system.[2]

First, a few definitions. **Meteorology** is the study of weather and

Merrilee Thomas/Tom Stack & Associates

The swirling winds of a tornado create positively charged ions.

weather forecasting. **Biometeorology** deals with the effects of environmental factors on living organisms. **Climate** is the average condition of the weather at a specific site, over a period of years. **Medical climatology** is concerned with the influences of natural climates on health. It deals with medical conditions that are either caused or aggravated by climatic elements. Many of these medical conditions were described in chapters 2-6, and include heat exhaustion, heatstroke, frostbite, hypothermia, barotrauma, decompression sickness, acute mountain sickness, headaches, and climatic aggravation of asthma and bronchitis.[3] Therefore, biometeorology is concerned with physiological responses, and medical climatology focuses on the interface of physiology and illness.

Weather systems and the climates of particular geographic regions consist of many geophysical factors, including temperature, wind velocity, humidity, barometric pressure, precipitation, positive or negative air ion density, and radiation (e.g., solar, infrared, ultraviolet). For centuries, scientists have theorized that weather, and especially changes in meteorological systems, affect human behavior, health, and performance. One of the most significant of the early systematic studies was conducted by the French sociologist Emile Durkheim, who evaluated the influence of temperature and climate on suicide.[4] He concluded that suicide death rates were *not* influenced by climate. However, this does not rule out the effects of atmospheric conditions on behavior. In order to resolve this issue, it is necessary to consider all of the elements that make up the physical environment and assess how they can jointly or independently influence behavior.

UNDERSTANDING AIR IONS

Centuries of folk lore and anecdotal reports suggest that some people suffer discomfort and pain from the slightest change of weather, while others respond only to severe changes in weather.[5] A thorough review of the research regarding the effects of short-term weather systems and climate on humans reveals that only a few well-designed studies have been published. For example, Muecher and Ungeheuer investigated reaction time, visual perception, job accident proneness, and visits to a medical dispensary during six different weather conditions.[6] They concluded that these four variables were affected by *changes* in the weather, especially by a dry warm wind that visits the Alpine regions of Europe in early spring and fall; this wind has been named *Foehn*, or "Witches' Wind," because of the deleterious effects it has on humans. There are numerous other winds of this type around the world, including the *Santa Ana* in California, the *Chinook* in western Canada and United States, the *Sirocco* in Italy, the *Sharav*, or *Hamsin*, of the Middle East, the *Mistral* in France, the *Thar* in India, and the *Zonda* in Argentina.[7,8]

These hot, dry, ground-level winds carry huge quantities of dust and generate enormous fields of static electricity and ions. **Ions** are formed when atoms gain or lose electrons (negatively charged particles), creating an imbalance in the number of charged particles in each. The positive ion concentrations in these storms may increase to 13,000 per cubic centimeter (cc). Over open land, there are normally 400 to 2000 total ions in every cc of air (1 cubic in. = 16.4 cc), and the ratio of + to − ions ranges from 4:1 to 1.3:1. Several other natural phenomena generate negative and positive ions (figure 7.1), on a smaller scale.

Figure 7.2 presents a striking example of the influence of a moving weather front on positive and negative air ions. These measurements were taken just as an afternoon rainstorm passed through Philadelphia, Pennsylvania. You will note that air ion concentrations rose rapidly during the 90 min prior to the storm, then fell continuously until the storm ended.

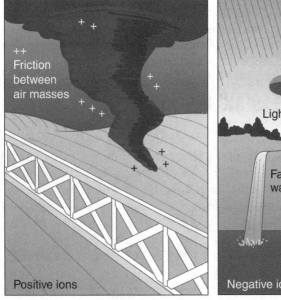
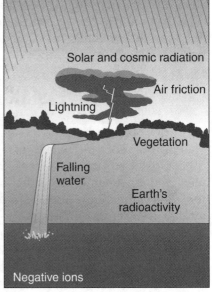

Figure 7.1 Natural phenomena that generate negative and positive ions.
Adapted from Wallach 1983.

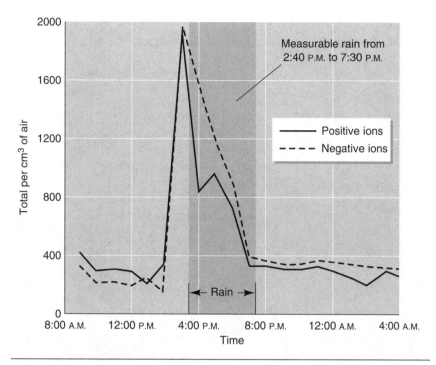

Figure 7.2 Total ion levels before, during, and after a rainstorm. This demonstrates the effect that a passing weather front has on both positive and negative ions.
Adapted from Davis and Speicher 1961.

YOUR BODY'S RESPONSES TO AIR IONS

Figure 7.3 illustrates the body systems that respond to changes in air ionization. These include the nervous, endocrine, respiratory, and immune systems. The text below describes the nature of these responses.

It has been shown that up to 30% of humans are sensitive to the high positive ion levels in the dry, warm winds just discussed.[8] For example, Robinson and Dirnfeld studied the Sharav, whose prominent features include a sudden rise in temperature, a drop in humidity, and an accompanying wind.[10] They noted that weather-sensitive individuals began to suffer just at the time the total air ion count rose via a disproportionate increase in the number of positive ions; this was 24–48 h before any other changes occurred in environmental parameters such as wind velocity or direction, temperature, solar radiation, and humidity.[8] Symptoms such as migraine headaches, limb swelling, asthma, heart palpitations, and digestive tract hyperactivity are typical in "sensitive" individuals.[7]

The Body's Responses to Stress

Nervous System

High positive ion levels in air:
- increase serotonin (a neurotransmitter)
- increase symptoms of depression and irritability
- decrease mental performance

Respiratory System

- air ions enter the blood through the lungs
- increased serotonin results in dry throat and airways
- positive air ions may induce asthma

Immune System

- increased serotonin results in itchy sinuses, an allergy-like response
- high levels of positive air ions increase inflammation

Endocrine System

- the endocrine response to air ions is described well by the body's general response to stress (figure 1.2)

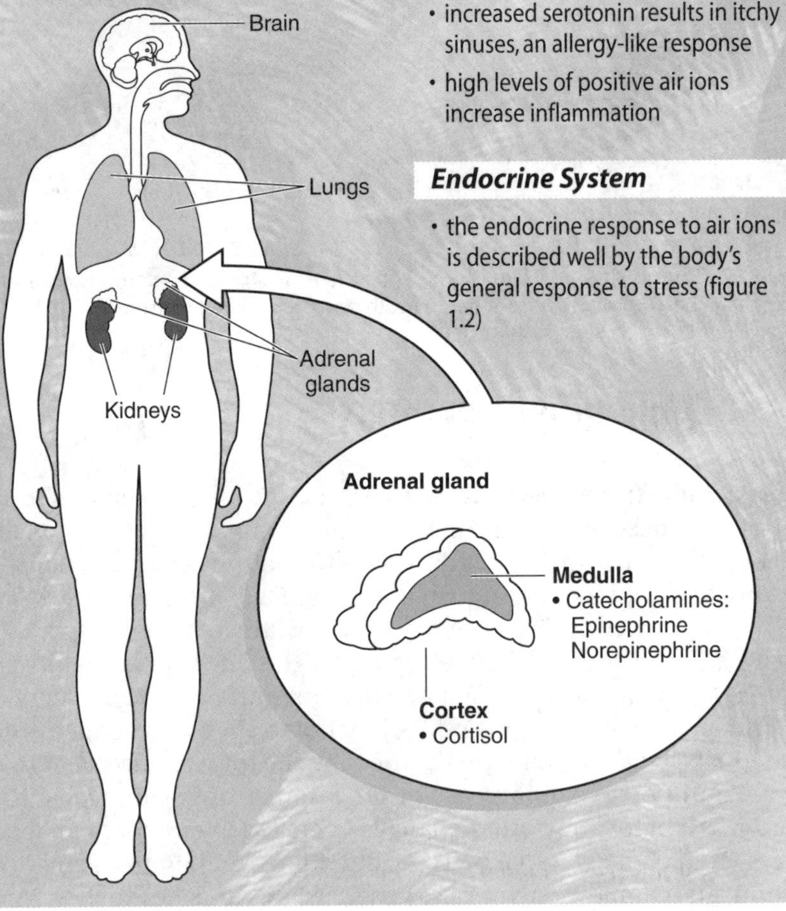

Brain

Lungs

Adrenal glands

Kidneys

Adrenal gland

Medulla
- Catecholamines:
  Epinephrine
  Norepinephrine

Cortex
- Cortisol

Figure 7.3 Systems of the body affected by air ionization.

Air Ions Effect on the the Brain

Robinson and Dirnfeld also named one cluster of symptoms the **"sero-tonin irritation syndrome."**[10] From chapter 1, you will recall that **serotonin** is a brain neurochemical that directly affects the pituitary and its secretion of the hormones ACTH and prolactin. This clinical name was assigned because the symptoms resembled hyperactivity of serotonin in the midbrain, and because patients excrete abnormally large amounts of serotonin in urine. A subsequent study showed that this illness can be successfully treated by inhalation of air containing large numbers of small negative ions, or by serotonin-blocking drugs.[11] In explaining these clinical observations, Sulman and colleagues concluded that the Sharav winds were stressors that imposed hormonal and metabolic changes on the human body.[12] And, in keeping with the principles of hormone action presented in chapter 1, additional observations were made regarding the Sharav winds: production of norepinephrine and epinephrine increased; responses differed with time of residence in the Middle East (i.e., short-term versus long-term); responses differed among individuals possibly because of differences in epinephrine secretion;[13] women are more sensitive to weather changes than men; and children (10-15%) are less sensitive that adults (20-30%).[14]

In a published summary, Doctor Sulman reviewed the nature of the human body's responses to the Sharav winds.[15] He identified two relevant responses to this harsh environmental stressor; both serve to diminish the HPA axis response to stressors.

• *Presensitivity to weather.* Air ions are positively or negatively charged oxygen (or water) molecules that reach a region 1-2 days before the arrival of a weather front. These ions enter the body via the alveoli of the lungs. Positive ions release an excess of serotonin, which provokes an irritation syndrome (see above) that inhibits the HPA hormonal axis.

• *Postsensitivity to weather.* Weather-sensitive patients react to hot-dry winds by releasing epinephrine. When exposure is prolonged, adrenal exhaustion can result (see above) and the body's response to the Sharav winds is inadequate because insufficient epinephrine is produced.

The existence of the serotonin irritation syndrome was supported by Jonathan Charry's doctoral dissertation, conducted at New York University in 1974.[11] It demonstrated that this syndrome involved a slower visual reaction time and increased fatigue in sensitive subjects when positive ions were presented in high concentrations.

Psychological Effects of Air Ions

The Sharav winds also induce psychological effects. This might be expected, if this weather system induces increased feelings of tension, behavioral dysfunction, and migraine headaches,[6, 12] as well as depression and irritability in weather-sensitive people.[16] For example, the Israeli researcher Rim compared the psychological test scores of two groups of subjects: those who tested on Sharav days (193 people) and those who tested on non-Sharav days (190 people), in the same months.[17] These individuals were candidates for either clerical or technological jobs. This study demonstrated that the Sharav winds caused an increase in neuroticism and extraversion, and a decrease in intelligence score, mechanical comprehension, and ability to follow instructions. The authors attributed these decrements in performance, in part, to the increased concentration of positive air ions found when Sharav winds are present.

Two years later, Rim published a follow-up study that involved psychological testing similar to that in the study above.[17] These tests were administered to 440 volunteers in a microclimate enriched with small negative ions that *reverse* the clinical symptoms associated with Sharav weather systems.[11] Of these subjects, 225 were tested on Sharav days and 215 were tested on non-Sharav days. Rim observed that people with specific personality traits were affected differently by the Sharav weather front (i.e., positive air ions) and negative air ion conditions. Introverts, for example, showed a decrease in short-term memory performance. Also, those who scored high on both neurotic behavior and extraversion showed the greatest improvement of responses due to negative air ions. These studies suggest that (a) the Sharav winds can alter psychological performance, and (b) humans are affected differently, perhaps because of their personality type (brain neurochemistry). Several other studies have shown, however, that air ionization had no effect on complex mental tasks.[12]

Physiological Effects of Air Ions

It is believed that small air ions (both + and –) enter the body through the respiratory tract. Both animal and human research studies have shown that excessive positive ionization of inhaled air reduced the motion of cilia in the windpipe and caused dry throats, headaches, and itchy or obstructed sinuses. Negative ionization resulted in none of these symptoms.[18] Despite logical skepticism about the effects of air ions on physiological function,[3] two intriguing studies provide biochemical evidence that a brain neurochemical is intimately involved.

These studies were prompted by the fact that the effects of serotonin (e.g., smooth muscle contraction, constriction of blood vessels, increased respiratory rates) were very similar to the effects attributed to positive ions. Investigators also hypothesized that negative ions would be able to counteract the effects of serotonin. In two separate experiments, drugs were used that have the opposite effects on the synthesis and removal of serotonin; reserpine depletes brain serotonin, whereas iproniazid acts to increase the supply of serotonin (mimicking the theoretical action of positive air ions).[19] With reserpine, tracheal tissue was resistant to positively ionized air. This supported the concept that the absence of serotonin was involved. When iproniazid was applied to tracheal tissue, the effects of positive air ions were duplicated (without actually using positive ions) and the influence of negative air ions was resisted. These findings illustrate that the brain neurotransmitter serotonin may be one part of the biological mechanism by which positive and negative air ions act in the body.[12] These results also might explain how weather-sensitive subjects are affected.

© Tom McCarthy/Unicorn Stock Photos

Strong, hot, ground-level winds generate enormous fields of static electricity and ions.

Several other experiments involving animals have shown that negative air ions exert a measurable anxiety-reducing effect on mice and rats exposed to stressors.[8] These trials employed sophisticated techniques to measure the responses of the endocrine and nervous systems to stressful situations. The hypothalamus, adrenal glands, brain metabolism, behavior, activity, eating, and adaptations to stressors were observed.[20]

Medical studies on humans strengthen these animal findings. You will recall that some individuals experience an irritation syndrome just before and during the Sharav winds. Physician Felix Sulman prescribed an anti-serotonin drug (sandomigran) for 80 Sharav-sensitive patients in 1971.[21] He observed that 70 of them lost all symptoms associated with serotonin excess, including migraine headaches, sleeplessness, irritability, and tension. Sulman and colleagues later demonstrated that Sharav-sensitive patients excreted larger amounts of serotonin in urine (50-500 μg units/24 h) than nonsensitive patients (0-50 μg units/24 h).[11]

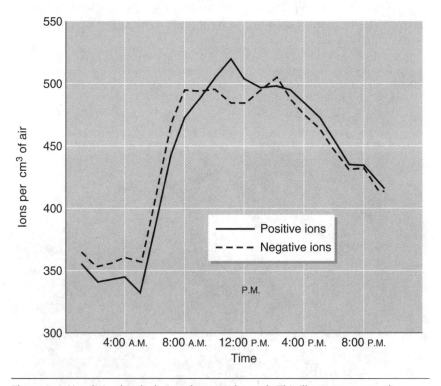

Figure 7.4 Hourly ion levels during clear weather only. This illustrates a natural environmental circadian rhythm in both positive and negative ions.
Adapted from Davis and Speicher 1961.

This suggests that there is some unknown characteristic of these weather-sensitive individuals that alters their biochemical balance under the influence of these meteorological stressors.[13]

Interestingly, even when weather systems are clear and calm, changes in air ion levels occur throughout the day. Figure 7.4 illustrates this phenomenon in a major city on the East Coast of the United States. Evidently, positive and negative ion levels peak between 8:00 A.M. and 4:00 P.M. Theoretically, this may be due to automotive and industrial pollution, or to the photochemical influence of sunlight, similar to ozone formation (see chapter 6). It is not clear from this study whether similar circadian changes in air ions occur in rural areas.

AIR IONS AND PHYSICAL PERFORMANCE

As early as 1961, Allan Frey of the General Electric Electronics Center at Cornell University published a review of the effects of atmospheric ions on behavior.[22] He concluded that reaction time in humans and the activity patterns of animals (i.e., movements per hour) were altered by changes in air ion concentrations and the ratio of positive to negative ions. Doctor Frey also summarized an obscure Russian study in which the work capacity of athletes was enhanced by inhaling negative ions for 25 days (15 min/day). These athletes increased their grip strength by 46% during this period. Also, after only 9 days of ionization, treadmill running endurance increased 60%.[23] However, Frey expressed concern over the experimental design of this study, stating that it may not have been properly controlled.

A few years after the above review was published, DeVries and Klafs reported on the role of breathing artificially ionized air on exercise performance.[24] Their working hypothesis was that breathing negatively ionized air will significantly improve endurance performance. Twenty-one male and 24 female physical education students participated in bench-stepping (50 cm bench, 36 steps/min) to exhaustion on four occasions. The inhaled air in each test differed in the following ways: (1) normal room air, (2) normal room air channeled through a mock ionization machine, (3) negatively ionized air from an ion generator, and (4) positively ionized air from an ion generator. The results indicated that artificially ionized air (either positive or negative) had no effect on endurance performance. In both this experiment and the study by Frey,[22] it is important to recognize that early ion generators were not as reliable as those in use today.[7]

More recent Israeli research reports have described the effects of a single exposure of negative air ions on two groups of male subjects.[25, 26]

The first group (experimental) received air that was neutral (221-256 ions/cc) on one occasion and air that was negatively charged (136,000-190,000 ions/cc) on another occasion. The second group (control) was exposed to the neutral environment twice. During 30 min exercise bouts in the heat (40°C, 104°F), negative air ion exposure resulted in significantly smaller rises in body temperature, heart rate, and perceived exertion, as well as significant improvements in work output. Several other physiological measurements (sweat rate, breathing rate, oxygen consumption, and blood lactic acid level) were not altered by inhalation of air ions. These findings showed that exposure to negative ions reduced physiological and mental strain during stressful exercise in a hot environment.

INTERPRETING AIR ION RESEARCH INTELLIGENTLY

The aforementioned investigations involving athletes represent the current state of research regarding air ions and physical performance.[27] Combined with other studies, some of which have shown both improvements in strength and endurance[23, 27] while others have shown no effect of air ions on physical performance,[28] this field can best be summarized by the word "controversial." The inconsistency probably results from the following factors:[27] (a) the concentration of air ions (as a unique stressor) must reach a critical threshold before internal homeostasis is disrupted and the SAM and HPA hormonal axes (see chapter 1) are called into play; (b) this critical threshold of air ions is unknown and probably is different for everyone;[8] (c) a person's response to air ionization probably differs at various times of the year and may differ daily, depending on the number and intensity of life's stressors, sleep, nutrition, and all other factors that make up good health; and (d) research studies involving air ions are very difficult to design and control. These facts remind us that we must consult reputable sources and scrutinize experimental designs before we naively accept statements about air ionization, biometeorology, or medical climatology as fact. For this reason, no recommendations regarding the ergogenic benefits or detrimental effects of air ionization can be made at this time. Further studies are required to clarify these matters. Nevertheless, the investigations conducted in Israel and intriguing anecdotal reports regarding Olympic athletes[27] suggest that future scientific and clinical research may lead to selected methods that will optimize human health and performance.

REFERENCES

1. Garmon, L. 1981. Something in the air. *Science News 120*: 364-365.

2. Spector, N.H.; Dolina, S.; Cornelissen, G.; Halberg, F.; Markovic, B.M.; & Jankovic, B.D. 1996. Neuroimmunomodulation: Neuroimmune interactions with the environment. In *Handbook of physiology*. Section 4: *Environmental physiology*. Vol. II, edited by M.J. Fregly & C.M. Blatteis, 1537-1550. New York: Oxford University Press.

3. Buettner, K.J.K., & Slonim, N.B. 1974. Biometeorology. In *Environmental physiology*, edited by N.B. Slonim, 42-60. St. Louis: Mosby.

4. Durkheim, E. 1897. *Suicide*. Translated by J. Spaulding & G. Simpson, 1951. New York: The Free Press, 104-122.

5. Irwin, T. 1976. *How weather and climate affect you*. New York: Public Affairs Committee, 1-28.

6. Muecher, H., & Ungeheuer, H. 1961. Meteorological influence on reaction time, flicker-fusion frequency, job accidents, and medical treatment. *Perceptual and Motor Skills 12*: 163-168.

7. Wallach, C. 1983. *The ion controversy: A scientific appraisal*. Australia: Belle Lumiere, 1-128. Lane Cove, New South Wales, Australia.

8. Krueger, A.P., & Reed, E.J. 1976. Biological impact of small air ions. *Science 193*: 1209-1213.

9. Davis, F.K., & Speicher, F.P. 1961. Natural ion levels in the City of Philadelphia. *Proceedings of the International Conference of Ionization of the Air*, 16-17 October, 1-16.

10. Robinson, N., & Dirnfeld, F.S. 1963. The ionization state of the atmosphere as a function of the meteorological elements of various sources of ions. *International Journal of Biometeorology 6*: 101-110.

11. Sulman, F.G.; Levy, D.; Lewy, A.; Pfeifer, Y.; Superstine, E.; & Tal, E. 1974. Air-ionometry of hot, dry desert winds (Sharav) and treatment with air ions of weather-sensitive subjects. *International Journal of Biometeorology 18*: 313-318.

12. Sulman, F.G.; Hirschman, N.; & Pfeifer, Y. 1964. Effect of hot dry desert winds (Sharav, Hamsin) on the metabolism of hormones and minerals. *Harokeach Haivri 10*: 401-404.

13. Charry, J.M. 1976. Meteorology and behavior: The effects of positive air ions on human performance, physiology, and mood, 1-374. PhD dissertation, New York University.

14. Sulman, F.G. 1982. *Short-and long-term changes in climate*. Boca Raton, FL: CRC Press, 65-136.

15. Sulman, F.G. 1983. Stress due to weather and climate. In *Selye's guide to stress research*. Vol. 2, edited by H. Selye, 10-21. New York: Scientific and Academic Editions.

16. Sulman, F.G.; Danon, A.; Pfeifer, Y.; Tal, E.; & Weller, C.P. 1970. Urinalysis of patients suffering from climatic heat stress. *International Journal of Biometeorology 14*: 45-53.

17. Rim, Y. 1975. Psychological test performance during climatic heat stress from desert winds. *International Journal of Biometeorology 19*: 37-40.

18. Soyka, F., & Edmonds, A. 1977. *The ion effect*. New York: Bantam Books, 48-80.

19. Krueger, A.P., & Smith, R.F. 1960. The biological mechanism of air ion action. *Journal of General Physiology 43*: 533-540.

20. Olivereau, J.M. 1971. Incidences psycho-physiologiques de l'ionisation atmosphèrique. Doctoral thesis. Paris: Université de Paris.

21. Sulman, F.G. 1971. Serotonin-migraine in climatic heat stress, its prophylaxis and treatment. *Proceedings of the International Headache Symposium 1*: 205-208.

22. Frey, A.H. 1961. Human behavior and atmospheric ions. *Psychological Review 68*: 225-228.

23. Vytchikova, M.A., & Minkh, A.A. 1959. On the use of aero-ionization in the practice of athletic medicine. *Physical Culture and Sport (USSR) 22*: 1-12.

24. DeVries, H.A., & Klafs, C.E. 1965. Ergogenic effects of breathing artificially ionized air. *Journal of Sports Medicine and Physical Fitness 5*: 7-12.

25. Inbar, O.; Rotshtein, A.; Dotan, R.; & Dalin, R. 1980. Effects of negative air ions on human performance. *Proceedings of the Seventh Annual IRA Symposium on Human Engineering and Quality of Work Life*, 314-343. Tel-Aviv, Israel: Wingate Institute.

26. Inbar, O.; Rotshtein, A.; Dalin, R.; Dotan, R.; & Sulman, F.G. 1982. The effects of negative air ions on various physiological functions during work in a hot environment. *International Journal of Biometeorology 22*: 153-163.

27. Armstrong, L.E. 1986. Air ions: An aid to performance? *National Strength and Conditioning Association Journal 8*: 35-37.

28. Davis, J.B. 1963. Review of scientific information on the effects of ionized air on human beings and animals. *Aerospace Medicine 34*: 35-42.

Biorhythmic Disturbances

chapter 8

> *Invisible rhythms underlie most of what we assume to be constant*
> *in ourselves and the world around us. Life is in continual flux, but*
> *the change is not chaotic. The rhythmic nature of earth life is,*
> *perhaps, its most usual yet overlooked property. Though we can*
> *neither see nor feel them, we are nevertheless surrounded by*
> *rhythms of gravity, electromagnetic fields, light waves, air pressure,*
> *and sound. Each day, as earth turns on its axis, we experience the*
> *alteration of light and darkness. The moon's revolution, too, pulls*
> *our atmosphere into a cycle of change.*
>
> *Night follows day. Seasons change. The tides ebb and flow. These*
> *various rhythms are also seen in animals and man. We, too, change,*
> *growing sleepy at night and restlessly active by day. We, too, exhibit*
> *the rhythmic undulations of our planet.*
>
> **—Gay Gaer Luce, 1971**[1]

You constantly encounter regular, rhythmic changes in numerous environmental factors. These rhythms cycle through peaks and valleys on a daily (daylight and darkness), monthly (the gravitational pull of the moon on ocean tides), and yearly (seasons, electromagnetic radiation) basis. Many normal responses of human organs and systems also are cyclic (biological rhythms) and may be influenced by cues from the environment, social interactions, and exercise. Further, these rhythms can be disrupted or desynchronized by travel across time zones and sleep loss.

Because biological rhythms are numerous and are affected by a host of variables, it may never be possible to manipulate them at will. This should not stop you, however, from attempting to understand them more fully and applying that knowledge to enhance your physical and mental performance. Recommendations are provided that will help you deal more effectively with travel across time zones and sleep loss.

THE RHYTHMS OF LIFE

Regular human **biological rhythms** were first discovered in 1657 by the scientist Sanctorius in Europe. Since that time, hundreds of biological rhythms have been identified in animals and plants, ranging from unicellular organisms to complex vertebrates. These rhythms are expressed as oscillations in physiological systems and last from minutes to months. All systems, organs, and tissues of the body exhibit biological rhythms (figure 8.1). **Circadian rhythms** last approximately 24 h (typical range: 20-28 h), and derive their name from the Latin phrase *circa diem*, which means "around a day," suggesting that they are synchronized with the rotation of the earth.

The following physiological processes and performance factors represent a few of the numerous human circadian rhythms: body temperature, heart rate, breathing rate, oxygen consumption, blood plasma volume, sweat rate, reaction time, neuromuscular coordination, flexibility of major joints, grip strength, muscular endurance, and physical work capacity.[4, 5, 6] Each of these increase and decrease imperceptibly in a rhythmic pattern. Psychological functions such as short-term memory, logical reasoning, mood state, vigor, and alertness also may show circadian oscillations.[1, 5] Figure 8.2 presents selected bodily processes that exhibit circadian rhythms in terms of their "peak hours" of function.

Control of Body Rhythms

The regular, rhythmic responses of organs and systems, as controlled by reverberating nerve circuits, may be lengthened or shortened by cues from the environment.[5, 6] An external cue, which resets the

The Body's Responses to Stress

Every body system has circadian, monthly, or annual rhythms that may be disturbed by crossing time zones and sleep loss:

- Nervous System
- Endocrine System
- Cardiovascular System
- Respiratory System
- Muscular System
- Immune System
- Urinary System
- Digestive System
- Reproductive System
- Integumentary (Skin)

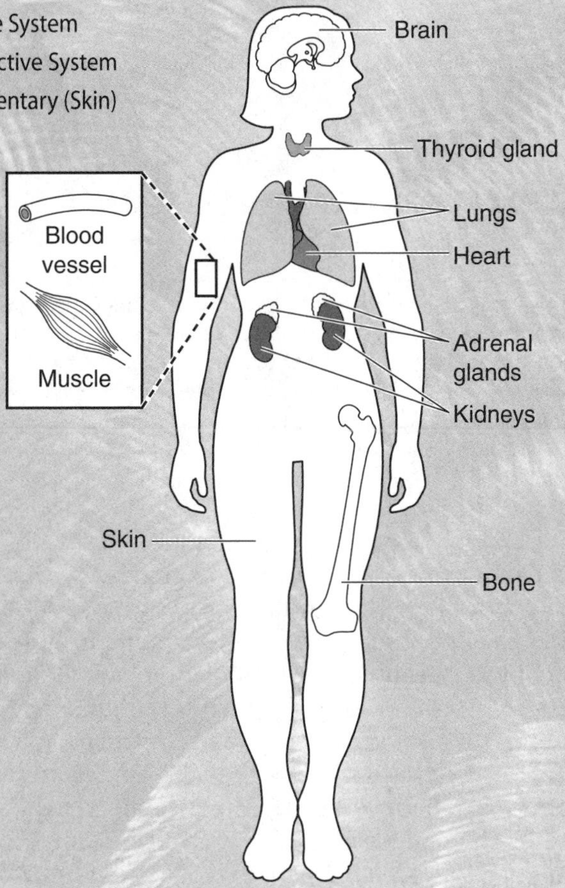

Brain

Thyroid gland

Lungs

Heart

Adrenal glands

Kidneys

Blood vessel

Muscle

Skin

Bone

Figure 8.1 Crossing time zones and sleep loss: effects on biological rhythms.

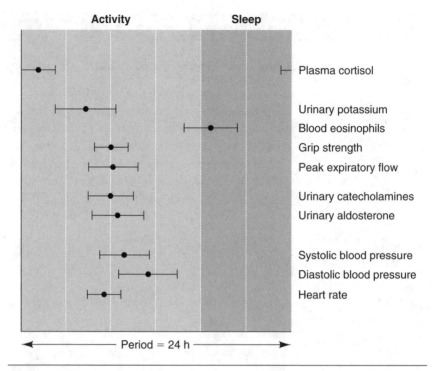

Figure 8.2 Human physiologic processes that exhibit circadian rhythms during daily activities (16 h) and periods of inactivity or sleep (8 h). Filled circles represent the time of peak function. Horizontal bars represent the range of measurements (95% confidence limits).

Adapted, by permission, from A. Reinberg and M.H. Smolensky, 1983, Introduction to chronobiology. In *Biological rhythms and medicine. Cellular, metabolic, physiopathologic, and pharmacologic aspects* (New York: Springer Verlag), 1-21.

duration of a biological rhythm, is named either a **zeitgeber** (a German word meaning "time giver"), **synchronizer**, or **entraining agent** by chronobiologists (i.e., those who study biological clocks). Besides light, which is the dominant external cue,[5] human zeitgebers include stressful situations,[6] group living,[7] shift work,[1, 7] meal timing,[8] specific dietary nutrients,[9] temperature, social interactions,[10] and exercise.[11]

The **hypothalamus**, which is the site of control for many biological responses (e.g., temperature, thirst, hunger, reproductive hormones), was for many years believed to be the primary "body clock." Presently, the suprachiasmatic nucleus (SCN) is considered by many to be the brain's master clock[12] because it regulates important neuroendocrine circadian rhythms such as ACTH, thyroid-stimulating hormone, prolactin, insulin, and glucagon.[3, 11, 13] However, because obliteration of the SCN in experimental animals leaves ultradian rhythms (those with a duration less than 24 h) intact, it is unlikely that the SCN acts alone to control biological rhythms.[3, 11] It is now believed that the body contains numerous circadian oscillators, at distinct sites, that are driven by signals such as sunrise (onset of light) and/or sunset (cessation of light). And, it is known that the microscopic structure and biochemical function of individual cells and cell components demonstrate 24 h variations. Examples of this include circadian rhythms in cell division (e.g., mitosis); mitochondrial enzyme activity (e.g., succinate dehydrogenase); lysosome number, size, and activity; Golgi apparatus activity; and the spatial arrangement of organelles (structures within each cell that have specific functions).[14] This also explains why isolated organs show circadian rhythmicity. For example, an independent circadian oscillator has been identified in the hamster retina, and a similar biological clock has been hypothesized in humans.[15]

Further, **genes** (sections of the DNA molecule in chromosomes that code for specific traits) are known to control at least one complex circadian behavior in mice (wheel running activity in the dark versus sleep during daylight hours).[11, 16] Genes that control other circadian rhythms have been discovered in hamsters, lizards, bacteria, mold, and fruit flies.[11, 17] Apparently, these genes produce peptides that regulate biological rhythms. Interestingly, periodic phenomena that are genetic in origin are not restricted to 24 h, 28 days, or 365 days. Biological rhythms of approximately 1 h, 90 min, 3 h, 7 days, and 30 days have been detected that have no correspondence to any known cosmic or environmental periodicities.[12, 18]

Illness and Medications

Susceptibility to diseases also may vary at different times of the day, month, or year (figure 8.3). For example, circadian variability has been observed in the onset of allergies, dermatitis, myocardial infarction, cerebral hemorrhage, as well as pain from dental caries, rheumatoid arthritis, and osteoarthritis of the hip and knee.[12] Monthly rhythms exist for the incidence of urticaria (hives related to immune system dysfunction) as well as bacterial and viral infections in women (figure 8.3). Both circadian and monthly rhythms have been observed for asthma, epilepsy, migraine headache, and bacterial/viral infections (see figure 8.3). Regular annual cycles also have been verified for upper respiratory infections, nonrespiratory infections (e.g., chicken pox, mumps, rubella, sexually transmitted diseases), and certain birth defects (e.g., congenital dislocation of the hip). These seasonal

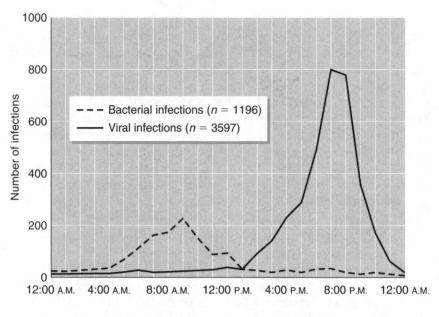

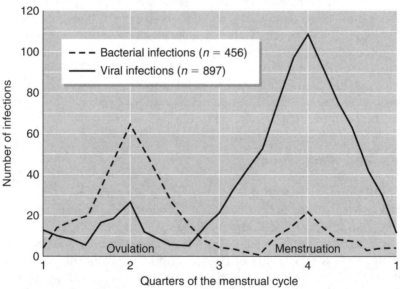

Figure 8.3 The onset of fever in women, signaling bacterial and viral infections. Top panel (circadian rhythm): Bacterial fevers are more common in the morning, whereas viral fevers occur more often in the evening. Bottom panel (circamensual rhythm): Bacterial infection is more likely around ovulation, whereas viral infection is more likely around menstruation.

Reprinted from Hejl 1977.

differences in the incidence of disease may be linked with the production and function of hormones. For example, there is a striking association of the annual rhythm of adrenal gland activity (which peaks in the spring season) with (a) the annual peak in the incidence of pneumonia and influenza, and (b) rhythms of immune status.[12]

Even the effectiveness of medications may be different at different points within a day or month.[20] Pharmacologic studies have shown this to be true for aspirin, acetaminophen, ethyl alcohol, diazepam, corticosteroids (e.g., prednisone, methylprednisolone), cardiovascular drugs (e.g., digitalis, atenolol, propranolol), antihistamines, diuretics, and antibiotics (e.g., erythromycin, ampicillin).[18, 20] This occurs, in part, because the amount of a medication that is absorbed into the bloodstream, the amount that is excreted, and the dynamic changes in the plasma concentration all vary.[18] In addition, some drugs (e.g., psychiatric medications such as lithium) actually may lengthen circadian and ultradian rhythms. Thus, it appears that *most* drugs may produce differential effects depending on the timing of administration.

Windows of Optimal Performance

Observations of athletes, who are required to perform optimally while under great duress and whose performance is easily measured (in terms of time, distance, strength, or points scored), provide valuable insights regarding human performance at different times of the day. Such studies have shown that the majority of performance-critical circadian rhythms are optimal in the afternoon.[4] Table 8.1, which was compiled from three sources, shows that many physiological and psychological factors peak between the times of 2:00 P.M. and 4:00 P.M. [5, 6, 8] You should realize, however, that this table presents an *average* time of day for each factor and that differences in circadian rhythms exist between athletes ("larks" are different from "owls" in epinephrine secretion, psychological mood, and activity patterns).[17] Also, the unique requirements of each sporting event or work task dictate how each factor in table 8.1 alters performance. For example, some activities require small muscle coordination and visual acuity more than others.

Coaches and supervisors can identify the optimal time of day for each athlete by keeping detailed records of previous performance during training sessions and competition. Also, the average athlete can determine his or her own peaks, valleys, and patterns of performance. This involves taking simple measurements of physical and

Table 8.1
Optimal Time of Day for Circadian Rhythms That Affect Athletic
Performance

Circadian rhythm	Time of day
Blood catecholamine level[6]	6:00 A.M.-10:00 A.M.
Blood cortisol level[6]	6:00 A.M.-10:00 A.M.
Short-term memory[6]	8:00 A.M.-1:00 P.M.
Speed and accuracy of motor performance[6, 8]	12:00 P.M.
Logical reasoning[6]	2:00 P.M.
Trunk flexibility[6]	2:00 P.M.
Vigor (self-rated)[6]	3:00 P.M.
Maximal breathing (ventilation) rate[6]	3:00 P.M.
Lowest fatigue during maximal exercise[6]	4:00 P.M.
Peak esophageal (core) temperature[5]	4:00 P.M.
Grip strength[6]	2:00 P.M.-6:00 P.M.
Maximal oxygen consumption[6]	3:00 P.M.-8:00 P.M.
Eye-hand tracking control[6]	8:00 P.M.

Adapted from Armstrong 1989.

psychological variables (e.g., pulse rate, blood pressure, addition speed, mood state, rating of vigor) several times a day.[21]

However, circadian rhythms should not be confused with the discredited "biorhythm" concept, which was popularized in the 1970s.[5, 6, 7] Involving an unsubstantiated 23-day physical cycle, a 28-day emotional cycle, and a 33-day emotional cycle, this concept resulted in books, charts, and calculators that predict "favorable" and "unfavorable" days in one's life. This discredited concept states that biorhythms are established at birth, do not change, and are identical for all persons with the same birth date.[22] In contrast, scientific studies have shown that true circadian rhythms may be altered by a variety of factors and vary greatly among individuals.[1, 5-7]

Travel Across Time Zones

The synchronization of many biological rhythms is generally believed to be beneficial for survival, because (a) coordination of circadian rhythms enhances biological function, and (b) organisms experience natural changes in electromagnetic fields, light, temperature, humidity, and air pressure regularly. Also, daily, monthly, and yearly rhythms are believed to be beneficial because it is more economical to be programmed to anticipate the cyclic demands of environmental synchronizer signals than to rely on immediate responses to one set of environmental conditions that is repeated every day and another set that is repeated every night.[12]

Travel across time zones severely desynchronizes biological rhythms.

© Jeff Greenberg/Unicorn Stock Photos

The significance of coordinating biological rhythms is summarized by Ehret and colleagues:[23]

A creature with all its systems in strong synchrony somehow learned how to "put it all together."... Such fortunate creatures are rewarded by functional proficiency and longevity. The opposite is true, as well. When you experience environmental, social, psychological, industrial, or athletic stressors that subtly disrupt biological rhythms, you become less efficient and effective because the systems of your body are not coordinated. A basic principle of chronobiology states that internal synchronization of biological rhythms is critical to good health and well-being.[11] In contrast, research suggests that desynchronization of circadian rhythms can have detrimental effects on physical and mental performance.[2] It is even possible that desynchronization of annual rhythms results in some of the mental disorders that have been associated with the change of seasons (e.g., seasonal affective disorder).[24] It is noteworthy that humans are the only species that regularly choose a lifestyle that leads to a disruption of the phasing of internal rhythms.[11]

Rapid travel across time zones (latitudinal or transmeridian air travel) provides one of the most potent desynchronizing agents known to science. The change in light-dark cycles and training/work schedules can modify virtually any biological rhythm, and results in what is commonly known as **jet lag**. This condition results when internal circadian rhythms (entrained to the time zone of departure) are "out of synch" with environmental cues at the time zone of arrival. Imagine the desynchronization that astronauts experience, as they are exposed to light and dark changes several times during each 24 h period!

When an athlete travels from the East Coast to the West Coast of the United States, 3 h are added to the length of his or her day, delaying normal circadian rhythms. However, each day that this athlete resides on the West Coast, the cycles of body temperature and other physiological variables shift approximately 1 h per day. Although these circadian rhythms eventually match the zeitgebers of the new time zone, after several days,[2] it is difficult to adapt to transmeridian travel that involves a difference of more than five time zones. In this case, at least 7 days should be allowed for resynchronization.[5] In contrast, following an eastward flight, a traveler adapts to the new time zone by *advances* in biological rhythms. In this instance, body temperature and other cycles reach their peaks and valleys approximately 1 h earlier each day.[2] This likely explains why travelers report that flights from west to east (shortened day) disturb circadian rhythms more than flights from east to west (lengthened day).[5, 7, 25] Adaptation is faster after westbound travel because the days are longer. (For example, during an 8 h jet flight from New York to Los Angeles, the amount of daylight experienced is 15 h going west but only 9 h going east, if sunrise and sunset occur at 7:00 A.M. and 7:00 P.M., respectively.) North to south, and south to north, flights have little impact on circadian rhythms.[7]

Athletes may require 2 to 10 days to recover from international, transmeridian travel. The U.S. Olympic Committee recognized this prior to the 1988 Summer Olympic Games in Seoul, Korea. Seoul lies seven time zones from Los Angeles and ten from New York. The USOC prepared a brochure titled "From the U.S. to Seoul—How to Avoid Jet Lag," which included guidelines to help athletes avoid desynchronization of biological rhythms. Table 8.2 presents estimates of the number of days required to fully resynchronize various circadian rhythms after air travel that involves west to east flights spanning six to nine time zones. This information was gleaned from three reputable sources.[5, 7, 26] Clearly, resynchronization does not occur rapidly.

Table 8.2
Time Required to Resynchronize Biological Rhythms Disrupted by Transmeridian Air Travel

Desynchronized biological rhythm	Time required to resynchronize (days)
Sleep/wake[7]	2-3
Esophageal (core) temperature[5, 7]	3-5
Cortisol secretion[7]	8-21
Eye-hand coordination[7]	1-5
Psychomotor performance[7]	1-5
Flight simulation[7] (simple tasks)	1-5
(complex tasks)	1-5
Arm strength[26]	1-5
Lift and carry task[26]	1-5
Sprint time[26]	1-5

Adapted from Armstrong 1989.

☒ Recommendations for Counteracting Jet Lag

Because biological rhythms are numerous and are affected by a host of variables, it may never be possible to manipulate them at will. This should not stop you, however, from attempting to understand them more fully and applying that knowledge to enhance your physical and mental performance. For example, if you plan to travel across time zones, a variety of recommendations have been published to help you either reduce desynchronization or speed resynchronization of circadian rhythms.[4] Please bear in mind, however, that these methods are still experimental.

- You should arrive several days prior to performing at distant sites (table 8.2).[2, 6] Shephard recommends that you allow at least 14 days for a major time zone difference.[5]

- When you arrive, adjust your activity and sleep schedule to the local time zone. Because light and dark are important zeitgebers, do not live or sleep in a room that has no windows.[6]

- Mild exercise enhances resynchronization of biological rhythms.[6] Intense exercise late in the day, however, will probably cause disturbed sleep.

- You may find that modifying your diet (namely, altering the ratio of protein and carbohydrate that you eat) will speed resynchronization. Subjective reports indicate that this technique is successful,[9] but controlled experiments have not been conducted to verify its value.[27]

- Caffeine may be used to help control sleepiness during daylight hours.[27] Under the guidance of a physician, judicious use of a sleep-inducing medication may help control sleep loss but does not resynchronize the circadian sleep cycle.[28] Alcohol should not be used to induce sleep because it disturbs sleep patterns.[27]

- Some authorities recommend carefully exposing the body to multiple zeitgebers, at specific times, to shift the body's circadian rhythms before traveling across time zones.[8] These synchronizers include exposure to light, fasting, shifting meal times, adjusting the carbohydrate-protein-fat ratio of meals, and consuming coffee or tea at specific times of the day.[2] Although the use of multiple zeitgebers by humans has been shown to reduce fatigue for several days,[29] the efficacy of specific strategies requires validation by controlled scientific experiments.

Some of these recommendations are supported by scientific evidence more strongly than others. Experimenting with these techniques, over several months, is the best approach because it is likely that each person will respond uniquely to transmeridian travel, and perhaps even differently from one trip to the next. One thing is certain. Resynchronization is a complex process. The best action may only minimize the effects of jet lag, not eliminate them completely.

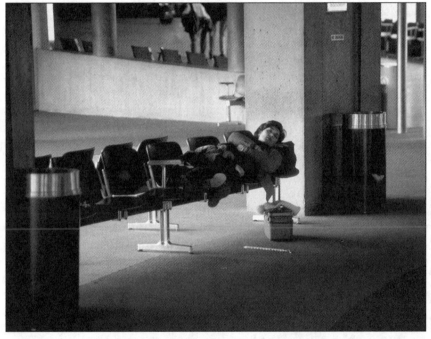

© Chris Boylan/Unicorn Stock Photos

Although traveling may disrupt sleep patterns, you *can* reduce the effects of jet lag.

SLEEP LOSS, HEALTH, AND PERFORMANCE

It is widely accepted that sleep is essential to good health. The condition known as "sleep loss" is difficult to define, but it begins when one gets considerably less than 7 h of sleep, the average amount needed by an adult each night. The following statistics, published in

1995, illustrate the wide-ranging effects that sleep loss may have on human health:[30]

- Missing 4 h of sleep reduces the immunological activity of cancer- and virus-fighting cells by 28%.
- Fatigue is the most common cause of death in truck accidents that are fatal to the driver.
- Twenty percent of all automobile drivers have fallen asleep at the wheel.
- Eighteen million Americans experience disruptive lapses in breathing (sleep apnea) during sleep, which often result in undiagnosed sleep loss.
- Sixty million Americans have frequent or chronic insomnia.

Interestingly, the risk of heatstroke (chapter 2) also increases when sleep loss is combined with moderate to severe exercise. In 1990, our research team reported that sleep loss and fatigue were the most prevalent risk factors observed among ten military heatstroke patients, even though air temperatures ranged only from 19 to 28°C (66-82°F).[31] This may be due to the fact that sweat sensitivity (i.e., the amount of sweat secreted per degree rise in core body temperature), evaporative heat loss, and dry heat dissipation are reduced by sleep deprivation.[32, 33]

Travel across time zones, or sustained activities, can modify human performance by desynchronizing sleep rhythms and causing sleep loss. This means that desynchronization and sleep loss may be viewed as two distinct effects on human performance that often coexist. Further, the physical fatigue and emotional stress associated with lengthy, sustained activities compound the effects of sleep loss. If athletes, laborers, and soldiers are active for a long period without sleep (e.g., during solo sailing, very long distance events, overtime work shifts, and military battles), normal zeitgebers such as light, meals, and social contacts are out of phase with the activity being performed.[5]

Three lengthy review articles have examined the effects of sleep loss on human exercise and cognitive performance, as reported in numerous scientific studies.[5, 34, 35] These reviews and others are summarized in table 8.3, which clearly demonstrates that some organs and systems are more sensitive to sleep loss than others. These studies also indicate that gross motor actions are less affected than cognitive function,[35] and activities that require vigilance or substantial cognitive effort deteriorate in proportion to the duration of sleep deprivation.[5] However, performance decrements are minor when sleep

Table 8.3

Summary of Research Studies That Evaluated Effects of Sleep Loss on Human Performance

Experimental outcome	Measurement	Sleep loss
Reduced	Creative thinking[36]	a
	Cognitive performance[37]	b
	Vigor[38]	c
	Arousal[5]	varied
	RPE[39]	60 h CON
	Sweat rate per °C rise in T_{es}[32]	33 h CON
	Evaporative and dry heat loss[33]	33 h CON
	Maximal/exhaustive exercise[40,41]	30-72 h CON
Increased	Confusion and fatigue[38]	c
	Plasma volume[42]	60 h CON
No change	Anger, tension, depression[38]	c
	Neuromuscular response time[43]	60 h CON
	Simple/practiced psychomotor tasks[5]	Varied
	Muscular strength[43]	60 h CON
	Exercise metabolism[44]	25-30 h CON
	Submaximal treadmill exercise[41]	30-72 h CON
	Cardiorespiratory responses[41]	30-72 h CON

Abbreviations: CON, consecutive; RPE, rating of perceived exertion; T_{es}, esophageal (core) temperature. Symbols: [a], 4-5 h of sleep per night during a 5-day tennis marathon; [b], 90 h CON, then 4 h of sleep per night during 6 days of military operations; [c], 3 h of sleep on 3 consecutive nights.

loss is less than 54 consecutive hours;[4] this is much greater than typical cases of jet lag-induced insomnia.

Thus, it is likely that the desynchronization of circadian rhythms will decrease mental and physical performance more often than sleep loss per se (table 8.3). Failure to allow adequate time for the resynchronization of circadian rhythms following transmeridian travel can have disastrous effects in major competitions.[5] This is relevant to the work of exercise scientists, fitness testing specialists, sports injury caregivers, military officers, industrial managers, and those who plan the travel of athletes or soldiers.[35]

Finally, there are wide differences among humans in their sensitivity to sleep deprivation.[5] These differences involve three considerations. First, it appears that the most responsive individuals tend to be anxious, neurotic, or extroverted.[45, 46] Second, a "morning person" and a "night person" may exhibit differences in epinephrine secretion, mood state, and activity patterns at different times of the day.[17] Third, aging affects the circadian rhythms of body temperature, hormone secretions, blood constituents, and the urinary excretion of metabolites.[47] The most consistent age-related differences involve a reduction of the amplitudes (i.e., a flattening) of these rhythms and an increase in the variability between individuals.[48]

Recommendations for Counteracting Sleep Loss

Significant sleep loss causes a decline in one's alertness and vigor and ability to perform mental tasks, and increases sleepiness and fatigue. This explains why athletes, laborers, and soldiers use two popular strategies to counteract sleep loss.

- Take a nap. Snoozing, after a lengthy period of sleep deprivation, causes a reversal of the effects described above. In general, a prophylactic nap of 2 to 4 hours has an effect on mental performance that is similar to consuming 150-300 mg of caffeine (1.5-3 cups of coffee).[50] Further, the beneficial effects of a nap and caffeine are additive. When these strategies are combined, they (a) are more effective than either strategy alone, and (b) allow adults to maintain alertness and mental performance near baseline levels (e.g., before sleep loss) during 24 h of continuous sleep loss.[51]

- Consume caffeine, either in coffee, soft drinks, or food. One cup of coffee may contain up to 100 mg of caffeine, depending on its mode of preparation. In the United States, an average adult coffee drinker consumes 200-400 mg of caffeine (2 to 4 cups of coffee) each day and 20-30 % of adults consume up to 600 mg/day.[49] Like taking a nap, caffeine reverses the detrimental effects of sleep loss on mental performance, alertness, and vigor.

- Athletes must consider one precaution. Although caffeine is safe as a food component at doses required to overcome sleep loss, and already exists in many foods, it is banned by the National Collegiate Athletic Association and the International Olympic Committee. This restriction exists because caffeine enhances exercise performance in endurance activities.[52] However, it is unlikely that caffeine enhances short-term, high-power events.[52, 53]

- Finally, a recent study suggests that physiological response may someday help athletes counteract sleep loss. This involves the fact that some people can simply decide to wake up in the morning, and are quite successful at waking within minutes of the time that they select before falling asleep. German scientists[54] have provided the first evidence of a biological basis for such an internal alarm clock—a rise in the hormone ACTH (and a subsequent rise in the stress hormone cortisol), just minutes prior to the ringing of an alarm clock, prepares the body for termination of a night's sleep. Knowing this, it is possible that athletes and laborers may someday avoid waking too early. Hopefully, future research will clarify this interesting phenomenon and provide practical methods that discourage the termination of sleep, such as not thinking about a specific wake-up time before going to bed at night.

REFERENCES

1. Luce, G.G. 1971. *Biological rhythms in human and animal physiology.* New York: Dover Publications, 1.

2. Haymes, E.M., & Wells, C.L. 1986. *Environment and human performance.* Champaign, IL: Human Kinetics, 121-122.

3. Reinberg, A., & Smolensky, M.H. 1983. Introduction to chronobiology. In *Biological rhythms and medicine. Cellular, metabolic, physiopathologic, and pharmacologic aspects*, 1-21. New York: Springer-Verlag.

4. Armstrong, L.E. 1989. Desynchronization of biological rhythms in athletes: Jet lag. *National Strength and Conditioning Association Journal 10*: 68-70.

5. Shephard, R.J. 1984. Sleep, biorhythms and human performance. *Sports Medicine 1*: 11-37.

6. Winget, C.M.; DeRoshia, C.W.; & Holley, D.C. 1985. Circadian rhythms and athletic performance. *Medicine & Science in Sports & Exercise 17*: 498-516.

7. Winget, C.M.; DeRoshia, C.W.; Markley, C.L.; & Holley, D.C. 1984. A review of human physiological and performance changes associated with desynchronosis of biological rhythms. *Aviation, Space, and Environmental Medicine 55*: 1085-1096.

8. Graeber, R.C.; Gatty, R.; Halberg, F.; & Levine H. 1978. *Human eating behavior: Preferences, consumptions, patterns, and biorhythms* (Technical Report No. TR-78/022). Natick, MA: U.S. Army Natick Research and Development Command.

9. Ehret, C.F.; Groh, K.R.; & Manert, J.C. 1980. Consideration of diet in alleviating jet lag. In *Chronobiology: Principles and applications to shifts and schedules*, edited by L.E. Scheving & F. Halberg, 393-402. Rockville, MD: Sijthoff and Noordhoff.

10. Fuller, C.A.; Sulzman, F.M.; & Moore-Ede, M.C. 1981. Shift work and the jet-lag syndrome: Conflicts between environmental and body-time. In *Biological rhythms, sleep and shift work*, 241-256. New York: SP Medical & Scientific Books.

11. Turek, F.W., & Van Reeth, O. 1996. Circadian rhythms. In *Handbook of physiology*. Section 4: *Environmental physiology*. Vol. II, edited by M.J. Fregly & C.M. Blatteis, 1329-1360. New York: Oxford University Press.

12. Nagai, K., & Nakagawa, H. 1992. Suprachiasmatic nucleus as a site of the circadian clock. In *Central regulation of energy metabolism with special reference to circadian rhythm*, 65-102. Boca Raton, FL: CRC Press.

13. Smolensky, M.H. 1983. Aspects of human chronopathology. In *Biological rhythms and medicine. Cellular, metabolic, physiopathologic, and pharmacologic aspects*, 131-209. New York: Springer-Verlag.

14. von Mayersbach, H. 1983. An overview of the chronobiology of cellular morphology. In *Biological rhythms and medicine. Cellular, metabolic, physiopathologic, and pharmacologic aspects*, 47-78. New York: Springer-Verlag.

15. Raloff, J. 1996. Eyes possess their own biological clocks. *Science News 149*: 245.

16. Anonymous. 1994. Ticktock, mice have a gene called clock. *Science News 145*: 319.

17. Barinagas, M. 1995. Shedding light on the ticking of internal timekeepers. *Science 267*: 1091-1092.

18. Reinberg, A. 1983. Clinical chronopharmacology. An experimental basis for chronotherapy. In *Biological rhythms and medicine. Cellular, metabolic, physiopathologic, and pharmacologic aspects*, 211-263. New York: Springer-Verlag.

19. Hejl, Z. 1977. Daily, lunar, yearly, and menstrual cycles and bacterial or viral infections in man. *Journal of Interdisciplinary Cycle Research 8*: 250-253.

20. Lemmer, B. 1988. Chronopharmacology of cardiovascular medications. In *Advanced in the biosciences*. Vol. 73, *Trends in chronobiology*, 219-242. Oxford, England: Pergamon Press.

21. Halberg, F.; Johnson, E.; Nelson, W.; Runge, W.; & Sothern, R. 1972. Autorhythmometry: Procedures for physiologic self-measurement and their analysis. *Physiology Teacher 1*: 1-11.

22. Thommen, G.S. 1973. *Is this your day? How biorhythms help determine your life cycles.* New York: Crown.

23. Ehret, C.F.; Groh, K.R.; & Manert, J.C. 1978. Circadian dyschronism and

chronotypic ecophilia as factors in aging and longevity. In *Aging and biological rhythms*, 185-214. New York: Plenum Press.

24. Rosenthal, N.E.; Sack, D.A.; Carpenter, C.J.; Parry, B.L.; Mendelson, B.W.; & Wehr, T.A. 1984. Seasonal affective disorder. *Archives of General Psychiatry 69*: 72-80.

25. Klein, K.E., & Wegmann, H.M. 1980. The effect of transmeridian and transequatorial air travel on psychological well-being and performance. In *Chronobiology: Principles and applications to shifts and schedules*, edited by L.E. Scheving & F. Halberg, 339-352. Rockville, MD: Sijthoff and Noordhoff.

26. Wright, J.E.; Vogel, J.A.; Sampson, J.B.; Knapik, J.J.; Patton, J.F.; & Daniels, W.L. 1983. Effects of travel across time zones (jet lag) on exercise capacity and performance. *Aviation, Space, and Environmental Medicine 54*: 132-137.

27. Aerospace Medical Association. 1996. Medical guidelines for air travel. *Aviation, Space, and Environmental Medicine 54*: B1-B16.

28. Donaldson, E., & Kennancy, D.J. 1991. Effects of tamazepam on sleep, performance, and rhythmic 6-sulpha-toxymelatonin and cortisol excretion after trans-meridian travel. *Aviation, Space, and Environmental Medicine 62*: 654-660.

29. Graeber, R.C. 1980. Recent studies relative to the airlifting of military units across time zones. In *Chronobiology: Principles and applications to shifts and schedules*, edited by L.E. Scheving & F. Halberg, 353-369. Rockville, MD: Sijthoff and Noordhoff.

30. Willensky, D. 1995. The best sleep statistics. *American Health 14*: 106.

31. Armstrong, L.E.; De Luca, J.P.; & Hubbard, R.W. 1990. Time course of recovery and heat acclimation ability of prior exertional heatstroke patients. *Medicine & Science in Sports & Exercise 22*: 36-48.

32. Kolka, M.A., & Stephenson, L.A. 1988. Exercise thermoregulation after prolonged wakefulness. *Journal of Applied Physiology 64*: 1575-1579.

33. Sawka, M.N.; Gonzalez, R.R.; & Pandolf, K.B. 1984. Effects of sleep deprivation on thermoregulation during exercise. *American Journal of Physiology 246*: R72-R77.

34. Home, J.A. 1978. A review of the biological effects of total sleep deprivation in man. *Biological Psychology 7*: 55-102.

35. Reilly, T. 1990. Human circadian rhythms and exercise. *Critical Reviews in Biomedical Engineering 18*: 165-180.

36. Horne, J.A. 1988. Sleep loss and "divergent" thinking ability. *Medicine & Science in Sport & Exercise 11*: 528-536.

37. Haslam, D.R. 1984. The military performance of soldiers in sustained operations. *Aviation, Space, and Environmental Medicine 55*: 216-221.

38. Reilly, T., & Piercy, M. 1994. The effect of partial sleep deprivation on weightlifting performance. *Ergonomics 37*: 107-115.

39. Montelpare, W.J.; Plyley, M.J.; & Shephard, R.J. 1992. Evaluating the influence of sleep deprivation upon circadian rhythms of exercise metabolism. *Canadian Journal of Sports Science 17*: 294-297.

40. Chen, H.I. 1991. Effects of 30-h sleep loss on cardiorespiratory functions at rest and in exercise. *Medicine & Science in Sport & Exercise 23*: 193-198.

41. VanHelder, T., & Radomski, M.W. 1989. Sleep deprivation and the effect on exercise performance. *Sports Medicine 7*: 235-247.

42. Goodman, J.; Radomski, M.; Hart, L.; Plyley, M.; & Shephard, R.J. 1989. Maximal aerobic exercise following prolonged sleep deprivation. *International Journal of Sports Medicine 10*: 419-423.

43. Symons, J.D.; Bell, D.G.; Pope, J.; VanHelder, T.; & Myles, W.S. 1988. Electromechanical response times and muscle strength after sleep deprivation. *Canadian Journal of Sports Sciences 13*: 225-230.

44. Hill, D.W.; Borden, D.O.; Darnaby, K.M.; & Hendricks, D.N. 1994. Aerobic and anaerobic contributions to exhaustive high-intensity exercise after sleep deprivation. *Journal of Sports Science 12*: 455-461.

45. Cappon, D., & Banks, R. 1960. Studies in perceptual distortion. *Archives of General Psychiatry 2*: 346-349.

46. Corcoran, D.W.J. 1962. Noise and loss of sleep. *Quarterly Journal of Experimental Psychology 14*: 178-182.

47. Brock, M.A. 1991. Chronobiology and aging. *Journal of the American Geriatric Society 39*: 74-79.

48. Atkinson, G., & Reilly T. 1996. Circadian variation in sports performance. *Sports Medicine 21*: 292-312.

49. Busto, U.; Bendayan, R.; & Sellers, E.M. 1989. Clinical pharmacokinetics of non-opiate abused drugs. *Clinical Pharmacokinetics 16*: 1-26.

50. Bonnet, M.H.; Gomez, S.; Wirth, O.; & Arand, D.L. 1995. The use of caffeine versus prophylactic naps in sustained performance. *Sleep 18*: 97-104.

51. Bonnet, M.H., & Arand, D.L. 1994. The use of prophylactic naps and caffeine to maintain performance during a continuous operation. *Ergonomics 37*: 1009-1020.

52. Tarnopolsky, M.A. 1994. Caffeine and endurance performance. *Sports Medicine 18*: 109-125.

53. Dodd, S.L.; Herb, R.A.; & Powers, S.K. 1993. Caffeine and exercise performance: An update. *Sports Medicine 15*: 14-23.

54. Born, J.; Hansen, K.; Marshal, L.; Molle, M.; & Fehm, H.L. 1999. Timing the end of nocturnal sleep [letter]. *Nature 397*: 29-30.

APPENDIX A

AMERICAN COLLEGE OF SPORTS MEDICINE POSITION STAND: EXERCISE AND FLUID REPLACEMENT

This pronouncement was written for the American College of Sports Medicine by: Victor A. Convertino, PhD, FACSM (Chair); Lawrence E. Armstrong, PhD, FACSM; Edward F. Coyle, PhD, FACSM; Gary W. Mack, PhD; Michael N. Sawka, PhD, FACSM; Leo C. Senay, Jr., PhD, FACSM; and W. Michael Sherman, PhD, FACSM.

SUMMARY

American College of Sports Medicine. Position Stand on Exercise and Fluid Replacement. *Med. Sci. Sports Exerc.,* Vol. 28, No. 1, pp. i–vii, 1996. It is the position of the American College of Sports Medicine that adequate fluid replacement helps maintain hydration and, therefore, promotes the health, safety, and optimal physical performance of individuals participating in regular physical activity. This position statement is based on a comprehensive review and interpretation of scientific literature concerning the influence of fluid replacement on exercise performance and the risk of thermal injury associated with dehydration and hyperthermia. Based on available evidence, the American College of Sports Medicine makes the following general recommendations on the amount and composition of fluid that should be ingested in preparation for, during, and after exercise or athletic competition:

1. It is recommended that individuals consume a nutritionally balanced diet and drink adequate fluids during the 24 h period before an event, especially during the period that includes the meal prior to exercise, to promote proper hydration before exercise or competition.

2. It is recommended that individuals drink about 500 ml (about 17 ounces) of fluid about 2 h before exercise to promote adequate hydration and allow time for excretion of excess ingested water.

3. *During* exercise, athletes should start drinking early and at regular intervals in an attempt to consume fluids at a rate sufficient to replace all the water lost through sweating (i.e., body weight loss), or consume the maximal amount that can be tolerated.

4. It is recommended that ingested fluids be cooler than ambient temperature [between 15° and 22°C (59° and 72°F)] and flavored to enhance palatability and promote fluid replacement. Fluids should be readily available and served in containers that allow adequate volumes to be ingested with ease and with minimal interruption of exercise.

5. Addition of proper amounts of carbohydrates and/or electrolytes to a fluid replacement solution is recommended for exercise events of duration greater than 1 h since it does not significantly impair water delivery to the body and may enhance performance. During exercise lasting less than 1 h, there is little evidence of physiological or physical performance differences between consuming a carbohydrate-electrolyte drink and plain water.

6. During intense exercise lasting longer than 1 h, it is recommended that carbohydrates be ingested at a rate of 30–60 g · h^{-1} to maintain oxidation of carbohydrates and delay fatigue. This rate of carbohydrate intake can be achieved without compromising fluid delivery by drinking 600–1200 ml · h^{-1} of solutions containing 4–8% carbohydrates (g · 100 ml^{-1}). The carbohydrates can be sugars (glucose or sucrose) or starch (e.g., maltodextrin).

7. Inclusion of sodium (0.5–0.7 g · L^{-1} of water) in the rehydration solution ingested during exercise lasting longer than 1 h is recommended since it may be advantageous in enhancing palatability, promoting fluid retention, and possibly preventing hyponatremia in certain individuals who drink excessive quantities of fluid. There is little physiological basis for the presence of sodium in an oral rehydration solution for enhancing intestinal water absorption as long as sodium is sufficiently available from the previous meal.

INTRODUCTION

Disturbances in body water and electrolyte balance can adversely affect cellular as well as systemic function, subsequently reducing the ability of humans to tolerate prolonged exercise. Water lost during exercise-induced sweating can lead to dehydration of both intracellular and extracellular fluid compartments of the body. Even a small amount of dehydration (1% body weight) can increase cardiovascular strain as indicated by a disproportionate elevation of heart rate during exercise, and limit the ability of the body to transfer heat from contracting muscles to the skin surface where heat can be dissipated to the environment. Therefore, consequences of body water deficits can increase the probability for impairing exercise performance and developing heat injury.

The specific aim of this position statement is to provide appropriate guidelines for fluid replacement that will help avoid or minimize the debilitating effects of water and electrolyte deficits on physiological function and exercise performance. These guidelines will also address the rationale for inclusion of carbohydrates and electrolytes in fluid replacement drinks.

HYDRATION BEFORE EXERCISE

Fluid replacement following exercise represents hydration prior to the next exercise bout. Any fluid deficit prior to exercise can potentially compromise thermoregulation during the next exercise session if adequate fluid replacement is not employed. Water loss from the body due to sweating is a function of the total thermal load that is related to the combined effects of exercise intensity and ambient conditions (temperature, humidity, wind speed).[62,87] In humans, sweating can exceed 30 g · min⁻¹ (1.8 kg · h⁻¹).[2,31] Water lost with sweating is derived from all fluid compartments of the body, including the blood (hypovolemia),[72] thus causing an increase in the concentration of electrolytes in the body fluids (hypertonicity).[85] People who begin exercise when hypohydrated with concomitant hypovolemia and hypertonicity display impaired ability to dissipate body heat during subsequent exercise.[26,28,61,85,86] They demonstrate a faster rise in body core temperature and greater cardiovascular strain.[28,34,82,83] Exercise performance of both short duration and high power output, as well as prolonged moderate intensity endurance activities, can be impaired when individuals begin exercise with the burden of a previously incurred fluid deficit,[1,83] an effect that is exaggerated when activity is performed in a hot environment.[81]

During exercise, humans typically drink insufficient volumes of fluid to offset sweat losses. This observation has been referred to as "voluntary dehydration."[33,77] Following a fluid volume deficit created by exercise, individuals ingest more fluid and retain a higher percentage of ingested fluid when electrolyte deficits are also replaced.[71] In fact, complete restoration of a fluid volume deficit cannot occur without electrolyte replacement (primarily sodium) in food or beverage.[39,89] Electrolytes, primarily sodium chloride, and to a lesser extent potassium, are lost in sweat during exercise. The concentration of Na^+ in sweat averages ~50 mmol $\cdot$ L^{-1} but can vary widely (20–100 mmol $\cdot$ L^{-1}) depending on the state of heat acclimation, diet, and hydration.[6] Despite knowing the typical electrolyte concentration of sweat, determination of a typical amount of total electrolyte loss during thermal or exercise stress is difficult because the amount and composition of sweat varies with exercise intensity and environmental conditions. The normal range of daily U.S. intake of sodium chloride (NaCl) is 4.6 to 12.8 g (~80–220 mmol) and potassium (K^+) is 2–4 g (50–100 mmol).[63] Exercise bouts that produce electrolyte losses in the range of normal daily dietary intake are easily replenished within 24 h following exercise and full rehydration is expected if adequate fluids are provided. When meals are consumed, adequate amounts of electrolytes are present so that the composition of the drink becomes unimportant. However, it is important that fluids be available during meal consumption since most persons rehydrate primarily during and after meals. In the absence of meals, more complete rehydration can be accomplished with fluids containing sodium than with plain water.[32,55,71]

To avoid or delay the detrimental effects of dehydration during exercise, individuals appear to benefit from fluid ingested prior to competition. For instance, water ingested 60 min before exercise will enhance thermoregulation and lower heart rate during exercise.[34,56] However, urine volume will increase as much as 4 times that measured without preexercise fluid intake. Pragmatically, ingestion of 400–600 ml of water 2 h before exercise should allow renal mechanisms sufficient time to regulate total body fluid volume and osmolality at optimal preexercise levels and help delay or avoid detrimental effects of dehydration during exercise.

FLUID REPLACEMENT DURING EXERCISE

Without adequate fluid replacement during prolonged exercise, rectal temperature and heart rate will become more elevated compared with a well-hydrated condition.[13,19,29,54] The most serious effect of dehydration resulting from the failure to replace fluids during exercise is impaired heat dissipation, which can elevate body core temperature to dangerously high levels (i.e., >40°C). Exercise-induced dehydration causes hypertonicity of body fluids and impairs skin blood flow[26,53,54,65] and has been associated with reduced sweat rate,[26,85] thus limiting evaporative heat loss, which accounts for more than 80% of heat loss in a hot-dry environment. Dehydration (i.e., 3% body weight loss) can also elicit significant reduction in cardiac output during exercise since a reduction in stroke volume can be greater than the increase in heart rate.[53,80] Since a net result of electrolyte and water imbalance associated with failure to adequately replace fluids during exercise is an increased rate of heat storage, dehydration induced by exercise presents a potential for the development of heat-related disorders,[24] including potentially life-threatening heat stroke. [88,92] It is therefore reasonable to surmise that fluid replacement that offsets dehydration and excessive elevation in body heat during exercise may be instrumental in reducing the risk of thermal injury.[37]

To minimize the potential for thermal injury, it is advocated that water losses due to sweating during exercise be replaced at a rate equal to the sweat rate.[5,19,66,73] Inadequate water intake can lead to premature exhaustion. During exercise, humans do not typically drink as much water as they sweat and, at best, voluntary drinking only replaces about two-thirds of the body water lost as sweat.[36] It is common for individuals to dehydrate by 2–6% of their body weight during exercise in the heat despite the availability of adequate amounts of fluid.[33,35,66,73] In many athletic events, the volume and frequency of fluid consumption may be limited by the rules of competition (e.g., number of rest periods or time outs) or their availability (e.g., spacing of aid stations along a race course).While large volumes of ingested fluids ($1 L \cdot h^{-1}$) are tolerated by exercising individuals in laboratory studies, field observations indicate that most participants drink sparingly during competition. For example, it is not uncommon for elite runners to ingest less than 200 ml of fluid during distance events in a cool environment lasting more than 2 h.[13,66] Actual rates of fluid ingestion are seldom more than 500 ml $\cdot$ h[-1] [66,68] and most athletes allow themselves to become dehydrated by 2–3 kg of body weight in sports such as running, cycling, and the triathlon. It is clear that perception of thirst, an imperfect index of the magnitude of fluid deficit, cannot be used to provide complete restoration of water lost by sweating. As such, individuals participating in prolonged intense exercise must rely on strategies such as monitoring body weight loss and ingesting volumes of fluid during exercise at a rate equal to that lost from sweating, i.e., body weight reduction, to ensure complete fluid replacement. This can be accomplished by ingesting beverages that enhance drinking at a rate of one pint of fluid per pound of body weight reduction. While gastrointestinal discomfort has been reported by individuals who have attempted to drink at rates equal to their sweat rates, especially in excess of 1 L · h⁻¹, [10,13,52,57,66] this response appears to be individual and there is no clear association between the volume of ingested fluid and symptoms of gastrointestinal distress. Further, failure to maintain hydration during exercise by drinking appropriate amounts of fluid may contribute to gastrointestinal symptoms.[64,76] Therefore, individuals should be encouraged to consume the maximal amount of fluids during exercise that can be tolerated without gastrointestinal discomfort up to a rate equal to that lost from sweating.

Enhancing palatability of an ingested fluid is one way of improving the match between fluid intake and sweat output. Water palatability is influenced by several factors including temperature and flavoring.[25,36] While most individuals prefer cool water, the preferred water temperature is influenced by cultural and learned behaviors. The most pleasurable water temperature during recovery from exercise was 5°C,[78] although when water was ingested in large quantities, a temperature of ~15–21°C was preferred.[9,36] Experiments have also demonstrated that voluntary fluid intake is enhanced if the fluid is flavored[25,36] and/or sweetened.[27] It is therefore reasonable to expect that the effect of flavoring and water temperature should increase fluid consumption during exercise, although there is insufficient evidence to support this hypothesis. In general, fluid replacement beverages that are sweetened (artificially or with sugars), flavored, and cooled to between 15° and 21°C should stimulate fluid intake.[9,25,36,78]

The rate at which fluid and electrolyte balance will be restored is also determined by the rate at which ingested fluid empties from the stomach and is absorbed from the intestine into the blood. The rate at which fluid leaves the stomach is dependent on a complex interaction of several factors, such as volume, temperature, and composition of the ingested fluid, and exercise intensity. The most important factor influencing gastric emptying is the fluid volume in the stomach.[52,68,75] However, the rate of gastric emptying of fluid is slowed proportionately with increasing glucose concentration above 8%.[15,38] When gastric fluid volume is maintained at 600 ml or more, most individuals can still empty more than 1000 ml · h⁻¹ when the fluids contain a 4–8% carbohydrate concentration.[19,68] Therefore, to promote gastric emptying, especially with the presence of 4–8% carbohydrate in the fluid, it is advantageous to maintain the largest volume of fluid that can be tolerated in the stomach during exercise (e.g., 400–600 ml). Mild to moderate exercise appears to have little or no effect on gastric emptying while heavy exercise at intensities greater than 80% of maximal capacity may slow gastric emptying.[12,15] Laboratory and field studies suggest that during prolonged exercise, frequent (every 15–20 min) consumption of moderate (150 ml) to large (350 ml) volumes of fluid is possible. Despite the apparent advantage of high gastric fluid volume for promoting gastric emptying, there should be some caution associated with maintaining high gastric fluid volume. People differ in their gastric emptying rates as well as their tolerance to gastric volumes, and it has not been determined if the ability to tolerate high gastric volumes can be improved by drinking during training. It is also unclear whether complaints of gastrointestinal symptoms by athletes during competition are a function of an unfamiliarity of exercising with a full stomach or because of delays in gastric emptying.[57] It is therefore recommended that individuals learn their tolerance limits for maintaining a high gastric fluid volume for various exercise intensities and durations.

Once ingested fluid moves into the intestine, water moves out of the intestine into the blood. Intestinal absorptive capacity is generally adequate to cope with even the most extreme demands;[30] and at intensities of exercise that can be sustained for more than 30 min, there appears to be little effect of exercise on intestinal function.[84] In fact, dehydration consequent to failure to replace fluids lost during exercise reduces the rate of gastric emptying,[64,76] supporting the rationale for early and continued drinking throughout exercise.

ELECTROLYTE AND CARBOHYDRATE REPLACEMENT DURING EXERCISE

There is little physiological basis for the presence of sodium in an oral rehydration solution for enhancing intestinal water absorption as long as sodium is sufficiently available in the gut from the previous meal or in the pancreatic secretions.[84] Inclusion of sodium (<50 mmol · L^{-1}) in fluid replacement drinks during exercise has not shown consistent improvements in retention of ingested fluid in the vascular compartment.[20,23,44,45] A primary rationale for electrolyte supplementation with fluid replacement drinks is, therefore, to replace electrolytes lost from sweating during exercise greater than 4–5 h in duration.[3] Normal plasma sodium concentration is 140 mmol · L^{-1}, making sweat (~50 mmol · L^{-1}) hypotonic relative to plasma. At a sweat rate of 1.5 L · h^{-1}, a total sodium deficit of 75 mmol · h^{-1} could occur during exercise. Drinking water can lower elevated plasma electrolyte concentrations back toward normal and restore sweating,[85,86] but complete restoration of the extracellular fluid compartment cannot be sustained without replacement of lost sodium.[39,70,89] In most cases, this can be accomplished by normal dietary intake.[63] If sodium enhances palatability, then its presence in a replacement solution may be justified because drinking can be maximized by improving taste qualities of the ingested fluid.[9,25]

The addition of carbohydrates to a fluid replacement solution can enhance intestinal absorption of water.[30,84] However, a primary role of ingesting carbohydrates in a fluid replacement beverage is to maintain blood glucose concentration and enhance carbohydrate oxidation during exercise that lasts longer than 1 h, especially when muscle glycogen is low.[11,14,17,18,50,60] As a result, fatigue can be delayed by carbohydrate ingestion during exercise of duration longer than 1 h, which normally causes fatigue without carbohydrate ingestion.[11] To maintain blood glucose levels during continuous moderate-to-high intensity exercise, carbohydrates should be ingested throughout exercise at a rate of 30–60 g $\cdot$ h^{-1}. These amounts of carbohydrates can be obtained while also replacing relatively large amounts of fluid if the concentration of carbohydrates is kept below 10% (g $\cdot$ 100 ml $^{-1}$ of fluid). For example, if the desired volume of ingestion is 600 – 1200 ml $\cdot$ h^{-1}, then the carbohydrate requirements can be met by drinking fluids with concentrations in the range of 4–8%.[19] With this procedure, both fluid and carbohydrate requirements can be met simultaneously during prolonged exercise. Solutions containing carbohydrate concentrations >10% will cause a net movement of fluid into the intestinal lumen because of their high osmolality, when such solutions are ingested during exercise. This can result in an effective loss of water from the vascular compartment and can exacerbate the effects of dehydration.[43]

Few investigators have examined the benefits of adding carbohydrates to water during exercise events lasting less than 1 h. Although preliminary data suggest a potential benefit for performance,[4,7,48] the mechanism is unclear. It would be premature to recommend drinking something other than water during exercise lasting less than 1 h. Generally, the inclusion of glucose, sucrose, and other complex carbohydrates in fluid replacement solutions has equal effectiveness in increasing exogenous carbohydrate oxidation, delaying fatigue, and improving performance.[11,16,79,90] However, fructose should not be the predominant carbohydrate because it is converted slowly to blood glucose (not readily oxidized),[41,42] which does not improve performance.[8] Furthermore, fructose may cause gastrointestinal distress.[59]

FLUID REPLACEMENT AND EXERCISE PERFORMANCE

Although the impact of fluid deficits on cardiovascular function and thermoregulation is evident, the extent to which exercise performance is altered by fluid replacement remains unclear. Although some data indicate that drinking improves the ability to perform short duration athletic events (1 h) in moderate climates,[7] other data suggest that this may not be the case.[40] It is likely that the effect of fluid replacement on performance may be most noticeable during exercise of duration greater than 1 h and/or at extreme ambient environments.

The addition of a small amount of sodium to rehydration fluids has little impact on time to exhaustion during mild prolonged (>4 h) exercise in the heat,[73] ability to complete 6 h of moderate exercise,[5] or capacity to perform during simulated time trials.[20,74] A sodium deficit, in combination with ingestion and retention of a large volume of fluid with little or no electrolytes, has led to low plasma sodium levels in a very few marathon or ultra-marathon athletes.[3,67] Hyponatremia (blood sodium concentration between 117 and 128 mmol $\cdot$ L^{-1}) has been observed in ultra-endurance athletes at the end of competition and is associated with disorientation, confusion, and in most cases, grand mal seizures.[67,69] One major rationale for inclusion of sodium in rehydration drinks is to avoid hyponatremia. To prevent development of this rare condition during prolonged (>4 h) exercise, electrolytes should be present in the fluid or food during and after exercise.

Maintenance of blood glucose concentrations is necessary for optimal exercise performance. To maintain blood glucose concentration during fatiguing exercise greater than 1 h (above 65% $\dot{V}O_2$max), carbohydrate ingestion is necessary.[11,49] Late in prolonged exercise, ingested carbohydrates become the main source of carbohydrate energy and can delay the onset of fatigue.[17,19,21,22,51,58] Data from field studies designed to test these concepts during athletic competition have not always demonstrated delayed onset of fatigue,[46,47,91] but the inability to control critical factors (such as environmental conditions, state of training, drinking volumes) make confirmation difficult. Inclusion of carbohydrates in a rehydration solution becomes more important for optimal performance as the duration of intense exercise exceeds 1 h.

CONCLUSION

The primary objective for replacing body fluid loss during exercise is to maintain normal hydration. One should consume adequate fluids during the 24 h period before an event and drink about 500 ml (about 17 ounces) of fluid about 2 h before exercise to promote adequate hydration and allow time for excretion of excess ingested water. To minimize risk of thermal injury and impairment of exercise performance during exercise, fluid replacement should attempt to equal fluid loss. At equal exercise intensity, the requirement for fluid replacement becomes greater with increased sweating during environmental thermal stress. During exercise lasting longer than 1 h, (a) carbohydrates should be added to the fluid replacement solution to maintain blood glucose concentration and delay the onset of fatigue, and (b) electrolytes (primarily NaCl) should be added to the fluid replacement solution to enhance palatability and reduce the probability for development of hyponatremia. During exercise, fluid and carbohydrate requirements can be met simultaneously by ingesting $600-1200$ ml $\cdot$ h $^{-1}$ of solutions containing 4–8% carbohydrate. During exercise greater than 1 h, approximately $0.5-0.7$ g of sodium per liter of water would be appropriate to replace that lost from sweating.

ACKNOWLEDGMENT

This pronouncement was reviewed for the American College of Sports Medicine by members-at-large, the Pronouncement Committee, and by: David L. Costill, PhD, FACSM; John E. Greenleaf, PhD, FACSM; Scott J. Montain, PhD; and Timothy D. Noakes, MD, FACSM.

REFERENCES

1. Armstrong, L. E., D. L. Costill, & W. J. Fink. 1985. Influence of diuretic-induced dehydration on competitive running performance. *Med. Sci. Sports Exerc. 17*: 456–461.

2. Armstrong, L. E., R. W. Hubbard, B. H. Jones, & J. J. Daniels. 1986. Preparing Alberto Salazar for the heat of the 1984 Olympic marathon. *Physician Sportsmed. 14*: 73–81.

3. Armstrong, L. E., W. C. Curtis, R. W. Hubbard, R. P. Francesconi, R. Moore, & E. W. Askew. 1993. Symptomatic hyponatremia during prolonged exercise in heat. *Med. Sci. Sports Exerc. 25*: 543–549.

4. Ball, T. C., S. Headley, & P. Vanderburgh. 1994. Carbohydrate-electrolyte replacement improves sprint capacity following 50 minutes of high-intensity cycling. *Med. Sci. Sports Exerc. 26:* S196, (abstract).

5. Barr, S. I., D. L. Costill, & W. J. Fink. 1991. Fluid replacement during prolonged exercise: Effects of water, saline, or no fluid. *Med. Sci. Sports Exerc. 23*: 811–817.

6. Bean, W. B. & L. W. Eichna. 1943. Performance in relationship to environmental temperature. Reactions of normal young men to simulated desert environment. *Fed. Proc. 2*: 144–158.

7. Below, P. R. & E. F. Coyle. 1995. Fluid and carbohydrate ingestion individually benefit intense exercise lasting one hour. *Med. Sci. Sports Exerc. 27*: 200–210.

8. Bjorkman, O., K. Sahlin, L. Hagenfeldt, & J. Wahren. 1984. Influence of glucose and fructose ingestion on the capacity for longterm exercise. *Clin. Physiol. 4*: 483–494.

9. Boulze, D., P. Montastruc, & M. Cabanac. 1983. Water intake, pleasure and water temperature in humans. *Physiol. Behav. 30:* 97–102.

10. Brouns, F., W. H. M. Saris, & N. J. Rehrer. 1987. Abdominal complaints and gastrointestinal function during long-lasting exercise. *Int. J. Sports Med. 8*: 175–189.

11. Coggan, A. R. & E. F. Coyle. 1991. Carbohydrate ingestion during prolonged exercise: Effects on metabolism and performance. E*xerc. Sport Sci. Rev. 19*: 1– 40.

12. Costill, D. L. 1990. Gastric emptying of fluids during exercise. In *Perspectives in exercise science and sports medicine, Vol. 3, Fluid homeostatsis during exercise,* edited by C. V. Gisolfi & D. R. Lamb, 97–128. Carmel, IN: Benchmark Press, Inc.

13. Costill, D. L., W. F. Krammer, & A. Fisher. 1970. Fluid ingestion during distance running. *Arch. Environ. Health 21:* 520–525.

14. Costill, D. L. & M. Hargreaves. 1992. Carbohydrate nutrition and fatigue. *Sports Med. 13*: 86–92.

15. Costill, D. L. & B. Saltin. 1974. Factors limiting gastric emptying during rest and exercise. *J. Appl. Physiol. 37*: 679–683.

16. Coyle, E. F. 1991. Timing and method of increased carbohydrate intake to cope with heavy training, competition and recovery. *J. Sports Sci. 9*: 29–52.

17. Coyle, E. F., A. R. Coggan, M. K. Hemmert, & J. L. Ivy. 1986. Muscle glycogen utilization during prolonged strenuous exercise when fed carbohydrate. *J. Appl. Physiol. 61*: 165–172.

18. Coyle, E. F., J. M. Hagberg, B. F. Hurley, W. H. Martin, A. A. Ehsani, & J. O. Holloszy. 1983. Carbohydrate feeding during prolonged strenuous exercise can delay fatigue. *J. Appl. Physiol. 55*: 30–235.

19. Coyle, E. F. & S. J. Montain. 1992. Benefits of fluid replacement with carbohydrate during exercise. *Med Sci. Sports Exerc. 24* (Suppl. 9): S324–S330.

20. Criswell, D., K. Renshler, S. K. Powers, R. Tulley, M. Cicalé, & K. Wheeler. 1992. Fluid replacement beverages and maintenance of plasma volume during exercise: Role of aldosterone and vasopressin. *Eur. J. Appl. Physiol. 65*: 445–51.

21. Davis, J. M., W. A. Burgess, C. A. Slentz, W. P. Bartoli, & R. R. Pate. 1988. Effects of ingesting 6% and 12% glucose/electrolyte beverages during prolonged intermittent cycling in the heat. *Eur. Appl Physiol. 57*: 563–569.

22. Davis, J. M., D. R. Lamb, R. R. Pate, C. A. Slentz, W. A. Burgess, & W. P. Bartoli. 1988. Carbohydrate-electrolyte drinks: Effects on endurance cycling in the heat. *Am. J. Clin. Nutr. 48*: 1023–1030.

23. Deuster, P. A., A. Singh, A. Hofmann, F. M. Moses, & G. C. Chrousos. 1992. Hormonal responses to ingesting water or a carbohydrate beverage during a 2 h run. *Med. Sci. Sports Exerc. 24*: 72–79.

24. Eichna, L. W., W. B. Bean, W. F. Ashe, & N. Nelson. 1945. Performance in relation to environmental temperature. Reactions of normal young men to hot, humid (simulated jungle) environment. *Bull. Johns Hopkins Hosp. 76*: 25–58.

25. Engell, D. & E. Hirsch. 1990. Environmental and sensory modulation of fluid intake in humans. In *Thirst: Physiological and psychological aspects*, edited by D. J. Ramsay and D. A. Booth, 382–402. Berlin: Springer-Verlag.

26. Fortney, S. M. 1984. Effect of hyperosmolality on control of blood flow and sweating. *J. Appl. Physiol. 57*: 1688–1695.

27. Fortney, S. M., E. R. Nadel, C. B. Wenger, & J. R. Bove. 1981. Effect of acute alterations of blood volume on circulatory performance in humans. *J. Appl. Physiol. 50*: 292–298.

28. Fortney, S. M., E. R. Nadel, C. B. Wenger, & J. R. Bove. 1981. Effect of blood volume on sweating rate and body fluids in exercising humans. *J. Appl. Physiol. 51*: 1594–1600.

29. Gilsolfi, C. V. & J. R. Copping. 1974. Thermal effects of prolonged treadmill exercise in the heat. *Med. Sci. Sports 6*: 108–113.

30. Gisolfi, C. V., R. W. Summers, & H. P. Schedl. 1990. Intestinal absorption of fluids during rest and exercise. In *Perspectives in exercise science and sports medicine, Vol. 3. Fluid homeostasis during exercise*, edited by C. V. Gisolfi & D. L. Lamb, 129–180. Carmel, IN: Benchmark Press, Inc.

31. Gisolfi, C. V., K. J. Spranrer, R. W. Summers, H. P. Schedel, & T. L. Bleiler. 1991. Effects of cycle exercise on intestinal absorption in humans. *J. Appl. Physiol. 71*: 2518–2527.

32. Gonzalez-Alonso, J., C. L. Heaps, & E. F. Coyle. 1992. Rehydration after exercise with common beverages and water. *Int. J. Sports Med. 13*: 399–406.

33. Greenleaf, J. E. & F. Sargent II. 1965. Voluntary dehydration in man. *J. Appl. Physiol. 20*: 719–724.

34. Greenleaf, J. E. & B. L. Castle. 1971. Exercise temperature regulation in man during hypohydration and hyperhydration. *J. Appl. Physiol. 30*: 847–853.

35. Greenleaf, J. E., P. J. Brock, L. C. Keil, & J. T. Morse. 1983. Drinking and water balance during exercise and heat acclimation. *J. Appl. Physiol. 54*: 414–419.

36. Hubbard, R. W., O. Maller, M. N. Sawka, R. N. Francesconi, L. Drolet, & A. J. Young. 1984. Voluntary dehydration and alliesthesia for water. *J. Appl. Physiol. 57*: 868–875.

37. Hubbard, R. W. & L. E. Armstrong. 1988. The heat illness: Biochemical, ultrastructural, and fluid-electrolyte considerations. In *Human performance physiology and environmental medicine at terrestrial extremes*, edited by. K.

B. Pandolf, M. N. Sawka, & R. R. Gonzalez, 305–360. Indianapolis: Benchmark Press, Inc.

38. Hunt, J. N. & M. T. Knox. 1969. Regulation of gastric emptying. In *Handbook of physiology,* Vol. IV, 1917–1935. Washington, DC: American Physiological Society.

39. Lassiter, W. E. 1990. Regulation of sodium chloride distribution within the extracellular space. In *The regulation of sodium and chloride balance*, edited by D. W. Seldin & G. Giebisch, 23–58. New York: Raven Press, Inc.

40. Levine, L., M. S. Rose, R. P. Francesconi, P. D. Neufer, & M. N. Sawka. 1991. Fluid replacement during sustained activity in the heat: Nutrient solution vs. water. *Aviat. Space Environ. Med. 62*: 559–564.

41. Massicotte, D., F. Perronnet, C. Allah, C. Hillaire-Marcel, M. Ledoux, & G. Brisson. 1986. Metabolic response to [^{13}C]glucose and [^{13}C] fructose ingestion during exercise. *J. Appl. Physiol. 61*: 1180–1184.

42. Massicotte, D., F. Perronnet, G. Brisson, K. Bakkouch, & C. Hillaire-Marcel. 1989. Oxidation of glucose polymer during exercise: Comparison of glucose and fructose. *J. Appl. Physiol. 66*: 179–183.

43. Maughan, R. J. 1985. Thermoregulation and fluid balance in marathon competition at low ambient temperature. *Int. J. Sports Med. 6*: 15–19.

44. Maughan, R. J., C. E. Fenn, M. Gleeson, & J. B. Leiper. 1987. Metabolic and circulatory responses to the ingestion of glucose polymers and glucose-electrolyte solutions during exercise in man. *Eur. J. Appl. Physiol. 56*: 356–362.

45. Maughan, R. J., C. E. Fenn, M. Gleeson, & J. B. Leiper. 1989. Effects of fluid, electrolyte, and substrate ingestion on endurance capacity. *Eur. J. Appl. Physiol. 58*: 481–486.

46. Millard-Stafford, M., P. B. Sparling, L. B. Rosskopf, B. T. Hinson, & L. J. Dicarlo. 1990. Carbohydrate-electrolyte replacement during a simulated triathlon in the heat. *Med. Sci. Sports Exerc. 22*: 621–628.

47. Millard-Stafford, M., P. B. Sparling, L. B. Rosskopf, & L. J. Dicarlo. 1992. Carbohydrate-electrolyte replacement improves distance running performance in the heat. *Med. Sci. Sports Exerc. 24*: 934–940.

48. Millard-Stafford, M., L. B. Rosskopf, T. K. Snow, & B. T. Hinson. 1994. Pre-exercise carbohydrate-electrolyte ingestion improves one-hour running performance in the heat. *Med. Sci. Sports Exerc. 26*: S196 (abstract).

49. Mitchell, J. B., D. L. Costill, J. A. Houmard, M. G. Flynn, W. J. Fink, & J. D. Beltz. 1988. Effects of carbohydrate ingestion on gastric emptying and exercise performance. *Med. Sci. Sports Exerc. 20*: 110–115.

50. Mitchell, J. B., D. L. Costill, J. A. Houmard, W. J. Fink, D. D. Pascoe, & D. R. Pearson. 1989. Influence of carbohydrate dosage on exercise performance and glycogen metabolism. *J. Appl. Physiol. 67*: 1843–1849.

51. Mitchell, J. B., D. L. Costill, J. A. Houmard, W. J. Fink, R. A. Roberys, & J. A. Davis. 1989. Gastric emptying: Influence of prolonged exercise and carbohydrate concentration. *Med. Sci. Sports Exerc. 21*: 269–274.

52. Mitchell, J. B. & K. W. Voss. 1991. The influence of volume of fluid ingested on gastric emptying and fluid balance during prolonged exercise. *Med. Sci. Sports Exerc. 23*: 314–319.

53. Montain, S. J. & E. F. Coyle. 1992. Fluid ingestion during exercise increases skin blood flow independent of increases in blood volume. *J. Appl. Physiol. 73*: 903–910.

54. Montain, S. J. & E. F. Coyle. 1992. The influence of graded dehydration on hyperthermia and cardiovascular drift during exercise. *J. Appl. Physiol. 73*: 1340–1350.

55. Morimoto, T., K. Mike, H. Nose, S. Yamada, K. Hirakawa, & C. Matsubara. 1981. Changes in body fluid and its composition during heavy sweating and effect of fluid and electrolyte replacement. *Jpn. J. Biometeorol. 18*: 31–39.

56. Moroff, S. V. & D. B. Bass. 1965. Effects of overhydration on man's physiological responses to work in the heat. *J. Appl. Physiol. 20*: 267–270.

57. Moses, F. M. 1990. The effect of exercise on the gastrointestinal tract. *Sports Med. 9*: 159–172.

58. Murray, R., D. E. Eddy, T. W. Murray, J. G. Seifert, G. L. Paul, & G. A. Halaby. 1987. The effect of fluid and carbohydrate feeding during intermittent cycling exercise. *Med. Sci. Sports Exerc. 19*: 597–604.

59. Murray, R., G. L. Paul, J. G. Seifert, D. E. Eddy, & G. A. Halaby. 1989. The effects of glucose, fructose, and sucrose ingestion during exercise. *Med. Sci. Sports Exerc. 21*: 275–282.

60. Murray, R., G. L. Paul, J. G. Seifert, & D. E. Eddy. 1991. Responses to varying rates of carbohydrate ingestion during exercise. *Med. Sci. Sports Exerc. 23*: 713–718.

61. Nadel, E. R., S. M. Fortney, & C. B. Wenger. 1980. Effect of hydration state on circulatory and thermal regulations. *J. Appl. Physiol. 49*: 715–721.

62. Nadel, E. R., C. B. Wenger, M. F. Roberts, J. A. J. Stolwijk, & E. Cafarelli. 1977. Physiological defenses against hyperthermia of exercise. *Ann. N. Y. Acad. Sci. 301*: 98–110.

63. National Research Council. 1989. *Recommended dietary allowances,* 10th ed., 250–255. Washington, DC: National Academy Press.

64. Neufer, P. D., A. J. Young, & M. N. Sawka. 1989. Gastric emptying during exercise: Effects of heat stress and hypohydration. *Eur. J. Appl. Physiol. 58*: 433–439.

65. Nishiyasu. T., X. Shi, G. W. Mack, & E. R. Nadel. 1991. Effect of hypovolemia on forearm vascular resistance control during exercise in the heat. *J. Appl. Physiol. 71*: 1382–1386.

66. Noakes, T. D. 1993. Fluid replacement during exercise. *Exerc. Sports Sci. Rev. 21*: 297–330.

67. Noakes, T. D., R. J. Norman, R. H. Buck, J. Godlonton, K. Stevenson, & D. Pittaway. 1990. The incidence of hyponatremia during prolonged ultraendurance exercise. *Med. Sci. Sports Exerc. 22*: 165–170.

68. Noakes, T. D., N. J. Rehrer, & R. J. Maughan. 1991. The importance of volume in regulating gastric emptying. *Med. Sci. Sports Exerc. 23*: 307–313.

69. Noakes, T. D., N. Goodwin, B. L. Rayner, T. Branken, & R. K. N. Taylor. 1985. Water intoxication: A possible complication during endurance exercise. *Med. Sci. Sports Exerc. 17*: 370–375.

70. Nose, H., M. Morita, T. Yawata, & T. Morimoto. 1986. Recovery of blood volume and osmolality after thermal dehydration in rats. *Am. J. Physiol. 251*: R492-R498.

71. Nose, H., G. W. Mack, X. Shi, & E. R. Nadel. 1988. Role of osmolality and plasma volume during rehydration in humans. *J. Appl. Physiol. 65*: 325–331.

72. Nose, H., G. W. Mack, X. Shi, & E. R. Nadel. 1988. Shift in body fluid compartments after dehydration in humans. *J. Appl. Physiol. 65*: 318–324.

73. Pitts, G. C., R. E. Johnson, & F. C. Consolazio. 1944. Work in the heat as affected by intake of water, salt, and glucose. *Am. J. Physiol. 142*: 253–259.

74. Powers, S. K., J. Lawler, S. Dodd, R. Tulley, G. Landry, & K. Wheeler. 1990. Fluid replacement drinks during high intensity exercise: Effects on minimizing exercise-induced disturbances in homeostasis. *Eur. J. Appl. Physiol. 60*: 54–60.

75. Rehrer, N. J. 1994. The maintenance of fluid balance during exercise. *Int. J. Sports Med. 15*: 122–125.

76. Rehrer, N. J., E. J. Beckers, F. Brouns, F. Ten Hoor, & W. H. M. Saris. 1990. Effects of dehydration on gastric emptying and gastrointestinal distress while running. *Med. Sci. Sports Exerc. 22*: 790–795.

77. Rothstein, A., E. F. Adolph, & J. H. Wills. 1947. Voluntary dehydration. In *Physiology of man in the desert,* edited by E. F. Aldolph, 254–270. New York: Interscience.

78. Sandick, B. L., D. B. Engell, & O. Maller. 1984. Perception of water temperature and effects for humans after exercise. *Physiol. Behav. 32*: 851–855.

79. Saris, W. H. M., B. H. Goodpaster, A. E. Jeukendrup, F. Brouns, D. Halliday, & A. J. M. Wagenmakers. 1993. Exogenous carbohydrate oxidation from different carbohydrate sources during exercise. *J. Appl. Physiol. 75*: 2168–2172.

80. Sawka, M. N., R. G. Knowlton, & J. B. Critz. 1979. Thermal and circulatory responses to repeated bouts of prolonged running. *Med. Sci. Sports 11*:177–180.

81. Sawka, M. N., R. P. Francesconi, A. J. Young, & K. B. Pandolf. 1984. Influence of hydration level and body fluids on exercise performance in the heat. *J.A.M.A. 252*: 1165–1169.

82. Sawka, M. N., A. J. Young, R. P. Francesconi, S. R. Muza, & K. B. Pandolf. 1985. Thermoregulatory and blood responses during exercise at graded hypohydration levels. *J. Appl. Physiol. 59*: 1394–1401.

83. Sawka, M. N. & K. B. Pandolf. 1990. Effects of body water loss on physiological function and exercise performance. In *Perspectives in exercise science and sports medicine,* Vol. 3. *Fluid homeostasis during exercise,* edited by C. V. Gisolfi and D. R. Lamb, 1–38. Carmel, IN: Benchmark Press, Inc.

84. Schedl, H. P., R. J. Maughan, and C. V. Gisolfi. 1994. Intestinal absorption during rest and exercise: Implication for formulating an oral rehydration solution (ORS). *Med. Sci. Sports Exerc. 26*: 267–280.

85. Senay, L. C., Jr. 1968. Relationship of evaporative rates to serum [Na⁺], [K⁺], and osmolarity in acute heat stress. *J. Appl. Physiol. 25*: 149–152.

86. Senay, L. C., Jr. 1979. Temperature regulation and hypohydration: A singular view. *J. Appl. Physiol. 47*: 1–7.

87. Shapiro, Y., K. B. Pandolf, & R. F. Goldman. 1982. Predicting sweat loss response to exercise, environment, and clothing. *Eur. J. Appl. Physiol. 48*: 83–96.

88. Sutton, J. R. 1990. Clinical implications of fluid imbalance. In *Perspectives in exercise science and sports medicine*, Vol. 3. *Fluid homeostasis during exercise*, edited by C. V. Gisolfi & D. R. Lamb, 425–455. Carmel, IN: Benchmark Press, Inc.

89. Takamata, A., G. W. Mack, C.M. Gillen, & E. R. Nadel. 1994. Sodium appetite, thirst, and body fluid regulation in humans during rehydration without sodium replacement. *Am. J. Physiol. (Regulatory Integrative Comp. Physiol.) 266*: R1493–R1502.

90. Wagenmakers, J. M., F. Brouns, W. H. M. Saris, & D. Halliday. 1993. Oxidation rates of orally ingested carbohydrate during prolonged exercise in men. *J. Appl. Physiol. 75:* 2774–2780.

91. Wells, C. L., T. A. Schrader, J. R. Stern, & G. S. Krahenbuhl. 1985. Physiological responses to a 20-mile run under three fluid replacement treatments. *Med. Sci. Sports Exerc. 17:* 364–369.

92. Wyndham, C. H. 1977. Heat stroke and hyperthermia in marathon runners. *Ann. N. Y. Acad. Sci. 301*: 128–138.

APPENDIX B

AMERICAN COLLEGE OF SPORTS MEDICINE POSITION STAND: HEAT AND COLD ILLNESSES DURING DISTANCE RUNNING

This pronouncement was written for the American College of Sports Medicine by: Lawrence E. Armstrong, PhD, FACSM; (Chair), Yoram Epstein, PhD; John E. Greenleaf, PhD, FACSM; Emily M. Haymes, PhD, FACSM; Roger W. Hubbard, PhD; William O. Roberts, MD, FACSM; and Paul D. Thompson, MD, FACSM.

SUMMARY

American College of Sports Medicine. Position Stand on Heat and Cold Illnesses During Distance Running. *Med. Sci. Sports Exerc.,* Vol. 28, No. 12, pp. i-x, 1996. Many recreational and elite runners participate in distance races each year. When these events are conducted in hot or cold conditions, the risk of environmental illness increases. However, exertional hyperthemia, hypothermia, dehydration, and other related problems may be minimized with pre-event education and preparation. This position stand provides recommendations for the medical director and other race officials in the following areas: scheduling; organizing personnel, facilities, supplies, equipment, and communication; providing competitor education; measuring environmental stress; providing fluids; and avoiding potential legal liabilities. This document also describes the predisposing conditions, recognition, and treatment of the four most common environmental illnesses: heat exhaustion, heatstroke, hypothermia, and frostbite. The objectives of this position stand are: 1) To educate distance running event officials and participants about the most common forms of environmental illness including predisposing conditions, warning signs, susceptibility, and incidence reduction. 2) To advise race officials of their legal responsibilities and potential liability with regard to event safety and injury prevention. 3) To recommend that race officials consult local weather archives and plan events at times likely to be of low environmental stress to minimize detrimental effects on participants.

4) To encourage race officials to warn participants about environmental stress on race day and its implications for heat and cold illness. 5) To inform race officials of preventive actions that may reduce debilitation and environmental illness. 6) To describe the personnel, equipment, and supplies necessary to reduce and treat cases of collapse and environmental illness.

INTRODUCTION

This document replaces the position stand titled *The Prevention of Thermal Injuries During Distance Running*.[4] It considers problems that may affect the extensive community of recreational joggers and elite athletes who participate in distance running events. It has been expanded to include heat exhaustion, heatstroke, hypothermia, and frostbite—the most common environmental illnesses during races.

Because physiological responses to exercise in stressful environments may vary among participants, and because the health status of participants varies from day to day, compliance with these recommendations will not guarantee protection from environmentally induced illnesses. Nevertheless, these recommendations should minimize the risk of exertional hyperthermia, hypothermia, dehydration, and resulting problems in distance running and other forms of continuous athletic activity such as bicycle, soccer, and triathlon competition.

Managing a large road race is a complex task that requires financial resources, a communication network, trained volunteers, and teamwork. Environmental extremes impose additional burdens on the organizational and medical systems. Therefore, it is the position of the American College of Sports Medicine that the following RECOMMENDATIONS be employed by race managers and medical directors of community events that involve prolonged or intense exercise in mild and stressful environments.

1. Race Organization

a. Distance races should be scheduled to avoid extremely hot and humid and very cold months. The local weather history should be consulted when scheduling an event. Organizers should be cautious of unseasonably hot or cold days in early spring and late fall because entrants may not be sufficiently acclimatized. The windchill index should be used to reschedule races on cold, windy days because flesh may freeze rapidly and cold injuries may result.

b. Summer events should be scheduled in the early morning or the evening to minimize solar radiation and air temperature. Winter events should be scheduled at midday to minimize the risk of cold injury.

c. The heat stress index should be measured at the site of the race because meteorological data from a distant weather station may vary considerably from local conditions.[66] The wet bulb globe temperature (WBGT) index is widely used in athletic and industrial settings (see Appendix A)[87]. If the WBGT index is above 28°C (82°F), or if the ambient dry bulb temperature is below –20°C (–4°F), consideration should be given to canceling the race or rescheduling it until less stressful conditions prevail. If the WBGT index is below 28°C, participants should be alerted to the risk of heat illness by using signs posted at the start of the race and at key positions along the race course (see "Postscript: Measurement of Environmental Stress," page 310)[61]. Also, race organizers should monitor changes in weather conditions. WBGT monitors can be purchased commercially, or figure 2.9 may be used to approximate the risk of racing in hot environments based on air temperature and relative humidity. These two measures are available from local meteorological stations and media weather reports, or can be measured with a sling psychrometer.

d. An adequate supply of fluid must be available before the start of the race, along the race course, and at the end of the event. Runners should be encouraged to replace their sweat losses or consume 150–300 ml (5.3–10.5 oz) every 15 min.[3] Sweat loss can be derived by calculating the difference between pre and postexercise body weight.

e. Cool or cold (ice) water immersion is the most effective means of cooling a collapsed hyperthermic runner.[25,48,49,59,88] Wetting runners externally by spraying or sponging during exercise in a hot environment is pleasurable but does not fully attenuate the rise in body core temperature.[14,88] Wetting the skin can result in effective cooling once exercise ceases.

f. Race officials should be aware of the warning signs of an impending collapse in both hot and cold environments and should warn runners to slow down or stop if they appear to be in difficulty.

g. Adequate traffic and crowd control must be maintained along the course at all times.

h. Radio communication or cellular telephones should connect various points on the course with an information processing center to coordinate emergency responses.

2. Medical Director

A sports medicine physician should work closely with the race director to enhance the safety and provide adequate medical care for all participants. The medical director should understand exercise physiology, interpretation of meteorological data, heat and cold illness prevention strategies, potential liability, and the treatment of medical problems associated with endurance events conducted in stressful environments.

3. Medical Support

a. Medical organization and responsibility: The medical director should alert local hospitals and ambulance services and make prior arrangements to care for casualties, including those with heat or cold injury. Medical personnel should have the authority to evaluate, examine, and stop runners who display signs of impending illness or collapse. Runners should be advised of this procedure prior to the event.

b. Medical facilities: Medical support staff and facilities must be available at the race site. The facilities should be staffed with personnel capable of instituting immediate and appropriate resuscitation measures. The equipment necessary to institute both cooling therapy (ice packs, child's wading pools filled with tap water or ice water, fans) and warming therapy (heaters, blankets, hot beverages) may be necessary at the same event. For example, medical personnel treated 12 cases of hyperthermia and 13 cases of hypothermia at an endurance triathlon involving 2300 competitors: air temperature was 85°F, water temperature was 58°F.[92]

4. Competitor Education

The physical training and knowledge of competitive runners and joggers has increased greatly, but race organizers must not assume that all participants are well prepared or informed about safety. Distributing this position stand before registration, publicizing the event in the media, and conducting clinics or seminars before events are valuable educational procedures.

a. All participants should be advised that the following conditions may exacerbate heat illness: obesity,[13,39,89] low degree of physical fitness,[30,63,79,83] dehydration,[23,34,69,83,84,95] lack of heat acclimatization,[31,51,89] a previous history of heat stroke,[82,89] sleep deprivation,[5] certain medications, including diuretics and antidepressants,[31] and sweat gland dysfunction or sunburn.[31] Illness 1 wk prior to an event should preclude participation,[32,96] especially those involving fever, respiratory tract infections, or diarrhea.[41,46]

b. Prepubescent children sweat less than adults and have lower heat tolerance.[11, 12]

c. Adequate training and fitness are important for full enjoyment of the event and will reduce the risk of heat illness and hypothermia.[33,64,67,85]

d. Prior training in the heat will promote heat acclimatization[6] and thereby reduce the risk of heat illness, especially if the training environment is warmer than that expected during a race.[5,51] Artificial heat acclimatization can be induced in cold conditions.[6]

e. Adequate fluid consumption before and during the race can reduce the risk of heat illness, including disorientation and irrational behavior, particularly in longer events such as a marathon.[23,34,95]

f. Excessive consumption of pure water or dilute fluid (i.e., up to 10 liters per 4 hours) during prolonged endurance events may lead to a harmful dilutional hyponatremia,[60] which may involve disorientation, confusion, and seizure or coma. The possibility of hyponatremia may be the best rationale for inclusion of sodium chloride in fluid replacement beverages.[3]

g. Participants should be advised of the early symptoms of heat illness, which may include clumsiness, stumbling, headache, nausea, dizziness, apathy, confusion, and impairment of consciousness.[41,86]

h. Participants should be advised of the early symptoms of hypothermia (slurred speech, ataxia, stumbling gait) and frostbite (numbness, burning, pain, paresthesia) on exposed skin.[36] Wet clothing, especially cotton, increases heat loss and the risk of hypothermia.[68]

i. Participants should be advised to choose a comfortable running speed and not to run faster than environmental conditions or their cardiorespiratory fitness warrant.[43,71,91]

j. It is helpful if novice runners exercise with a partner, each being responsible for the other's well-being.[71]

5. Responsibilities and Potential Liability

The sponsors and directors of an endurance event are reasonably safe from liability due to injury if they avoid gross negligence and willful misconduct, carefully inform the participants of hazards, and have them sign waivers before the race.[78] However, a waiver signed by a participant does not totally absolve race organizers of moral and/or legal responsibility. It is recommended that race sponsors and directors: 1) minimize hazards and make safety the first concern; 2) describe inherent hazards (e.g., potential course hazards, traffic control weather conditions) in the race application; 3) require all entrants to sign a waiver; 4) retain waivers and records for 3 yr; 5) warn runners of the predisposing factors and symptoms of environmental illness; 6) provide all advertised support services; 7) legally incorporate the race or organizations involved; and 8) purchase liability insurance.[18,78,80]

Race directors should investigate local laws regarding Good Samaritan action. In some states physicians who do not accept remuneration may be classified as Good Samaritans. Race liability insurance may not cover physicians,[78] therefore the malpractice insurance policy of each participating physician should be evaluated to determine if it covers services rendered at the race.

Medical and race directors should postpone, reschedule, or cancel a race if environmental conditions warrant, even though runners and trained volunteers arrive at the site and financial sponsorship has been provided. Runners may not have adequate experience to make the decision not to compete; their safety must be considered. Downgrading the race to a "fun run" does not absolve race supervisors from their responsibility or decrease the risk to participants.[15,66]

BACKGROUND FOR THIS POSITION STAND

Dehydration is common during prolonged endurance events in both cold and hot environmental conditions because the average participant loses 0.5–1.5 quarts (0.47–1.42 liters) of sweat·h^{-1}, and fluid replacement is usually insufficient.[2,42,69] Runners may experience hyperthermia [body core temperature above 39°C (102.2°F)] or hypothermia [body core temperature below 35°C (95°F)], depending on the environmental conditions, caloric intake, fluid consumption, and clothing worn. Hyperthermia is a potential problem in warm and hot weather races when the body's rate of heat production is greater than its heat dissipation.[2] Indeed, on extremely hot days, it is possible that up to 50% of the participants may require treatment for heat-related illnesses such as heat exhaustion and heatstroke.[1,66] Hypothermia is more likely to occur in cold or cool-windy conditions. Scanty clothing may provide inadequate protection from such environments, particularly near the end of a long race when running speed and heat production are reduced. Frostbite can occur in low air temperature and especially when combined with high wind speed. The race and medical directors should anticipate the above medical problems and be capable of responding to a large number of patients with adequate facilities, supplies, and support staff. The four most common heat and cold illnesses during distance running are heat exhaustion, heatstroke, hypothermia, and frostbite.

1. Heat Exhaustion

Body sweat loss can be significant in summer endurance races and may result in a body water deficit of 6–10% of body weight.[41,95] Such dehydration will reduce the ability to exercise in the heat because decreases in circulating blood volume, blood pressure, sweat production, and skin blood flow all inhibit heat loss[41,81] and predispose the runner to heat exhaustion or the more dangerous hyperthermia and exertional heatstroke.[41,66]

Heat exhaustion, typically the most common heat illness among athletes, is defined as the inability to continue exercise in the heat.[7] It represents a failure of the cardiovascular responses to workload, high external temperature, and dehydration.[16,41,42] Heat exhaustion has no known chronic, harmful effects. Symptoms may include headache, extreme weakness, dizziness, vertigo, "heat sensations" on the head or neck, heat cramps, chills, "goose flesh" ("goose bumps"), vomiting, nausea, and irritability.[41,42] Hyperventilation, muscular incoordination, agitation, impaired judgment, and confusion also may be seen. Heat syncope (fainting) may or may not accompany heat exhaustion.[41] The onset of heat exhaustion symptoms is usually sudden and the duration of collapse brief. During the acute stage of heat exhaustion, the patient looks ashen-gray, the blood pressure is low, and the pulse rate is elevated. Hyperthermia may add to the symptoms of heat exhaustion, even on relatively cool days.[20,22,30,37,38,43,62,90]

Although it is improbable that all heat exhaustion cases can be avoided, the most susceptible individuals are those who either exert themselves at or near their maximal capacities, are dehydrated, not physically fit, and not acclimatized to exercise in the heat. It is imperative that runners be adequately rested, fed, hydrated, and acclimatized;[7] they should drink ample fluids before, during, and after exercise.[3] Also, repeated bouts of exercise in the heat (heat acclimatization) reduce the incidence of both heat exhaustion and heat syncope. Heat acclimatization can best be accomplished by gradually increasing the duration and intensity of exercise training during the initial 10–14 days of heat exposure.[6]

Oral rehydration is preferred for heat exhaustion patients who are conscious, coherent, and without vomiting or diarrhea. Intravenous (IV) fluid administration facilitates rapid recovery.[42,57] Although a variety of IV solutions have been used at races,[42] a 5% dextrose sugar in either 0.45% saline (NaCl) or 0.9% NaCl are the most common.[1] Runners may require up to 4 L of IV fluid if severely dehydrated.[57]

2. Exertional Heatstroke

Heat production, mainly from muscles, during intense exercise is 15–20 times greater than at rest, and is sufficient to raise body core temperature by 1°C (1.8°F) each 5 min without thermoregulatory (heat loss) adjustments.[56] When the rate of heat production exceeds that of heat loss for a sufficient period of time, severe hyperthermia occurs.

Heatstroke is the most serious of the syndromes associated with excess body heat. It is defined as a condition in which body temperature is elevated to a level that causes damage to the body's tissues, giving rise to a characteristic clinical and pathological syndrome affecting multiple organs.[32,83] After races, adult core (rectal) temperatures above 40.6°C (105.1°F) have been reported in conscious runners,[24,52,69,74,77] and 42–43°C (107.6–109.4°F) in collapsed runners.[72–74,86,90] Sweating is usually present in runners who experience exertional heatstroke.[87]

Strenuous physical exercise in a hot environment has been notorious as the cause of heatstroke, but heatstroke also has been observed in cool-to-moderate [13–28°C (55–82°F)] environments,[5,32,74] suggesting variations in individual susceptibility.[5,31,32] Skin disease, sunburn, dehydration, alcohol or drug use/abuse, obesity, sleep loss, poor physical fitness, lack of heat acclimatization, advanced age, and a previous heat injury all have been theoretically linked to increased risk of heatstroke.[5,31,51,84] The risk of heatstroke is reduced if runners are well-hydrated, well-fed, rested, and acclimatized. Runners should not exercise if they have a concurrent illness, respiratory infection, diarrhea, vomiting, or fever.[5,7,46] For example, a study of 179 heat casualties at a 14-km race showed that 23% reported a recent gastrointestinal or respiratory illness,[70] whereas a study of 10 military heatstroke patients reported that three had a fever or disease and six recalled at least one warning sign of impending illness at the time of their heatstroke.[5]

Appropriate fluid ingestion before and during prolonged running can minimize dehydration and reduce the rate of increase in body core temperature.[24,34] However, excessive hyperthermia may occur in the absence of significant dehydration, especially in races of less than 10 km, because the fast pace generates greater metabolic heat.[90]

The mortality rate and organ damage due to heatstroke are proportional to the length of time between core temperature elevation and initiation of cooling therapy.[5,26] Therefore, prompt recognition and cooling are essential.[1,5,22,42,48,51,62,74,83] A measurement of deep body temperature is vital to the diagnosis, and a rectal temperature should be measured in any casualty suspected of having heat illness or hypothermia. Ear (tympanic), oral, or axillary measurements are spuriously affected by peripheral (skin) and environmental temperatures and should not be used after exercise.[8,75,76] When cooling is initiated rapidly, most heatstroke patients recover fully with normal psychological status,[79] muscle energy metabolism,[65] heat acclimatization, temperature regulation, electrolyte balance, sweat gland function, and blood constituents.[5]

Many whole-body cooling techniques have been used to treat exertional heatstroke, including water immersion, application of wet towels or sheets, warm air spray, helicopter downdraft, and ice packs to the neck, underarm, and groin areas. There is disagreement as to which modality provides the most efficient cooling,[7,47,97] because several methods have been used successfully. However, the fastest whole-body cooling rates [25,48,49,59,88] and the lowest mortality rates[25] have been observed during cool and cold water immersion. Whichever modality is utilized it should be simple and safe, provide great cooling power, and should not restrict other forms of therapy (e.g., cardiopulmonary resuscitation, defibrillation, IV cannulation). The advantages and disadvantages of various cooling techniques have been discussed.[47,75,97]

Heatstroke is regarded as a medical emergency that might be fatal if not immediately diagnosed and properly treated. Early diagnosis is of utmost importance and time-consuming investigation should be postponed until body temperature is corrected and the patient is evacuated to a nearby medical facility that is aware of such conditions.

3. Hypothermia

Hypothermia [body core temperature below 36°C (97°F)] occurs when heat loss is greater than metabolic heat production.[94] Early signs and symptoms of hypothermia include shivering, euphoria, confusion, and behavior similar to intoxication. Lethargy, muscular weakness, disorientation, hallucinations, depression, or combative behavior may occur as core temperature continues to fall. If body core temperature falls below 31.1°C (88°F), shivering may stop and the patient will become progressively delirious, uncoordinated, and eventually comatose if treatment is not provided.[10]

During cool or cold weather marathons, the most common illnesses are hypothermia, exhaustion, and dehydration. The most common medical complaints are weakness, shivering, lethargy, slurred speech, dizziness, diarrhea, and thirst.[1,45] Runner complaints of feeling hot or cold do not always agree with changes in rectal temperature.[74] Dehydration is common in cool weather.[1,45] Runners should attempt to replace fluids at a rate that matches their sweat and urine losses. Cases of hypothermia also occur in spring and fall because weather conditions change rapidly and runners wear inappropriate clothing that becomes sweat-soaked during training or competition.[19]

Hypothermia may occur during races, for example when distance runners complete the second half of the event more slowly than the first half.[54] Evaporative and radiative cooling increase because wet skin (from sweat, rain, or snow) and clothing are exposed to higher wind speed at a time when metabolic heat production decreases. Hypothermia also occurs after a race, when the temperature gradient between the body surface and the environment is high. Subfreezing ambient temperatures need not be present and hypothermia may develop even when the air temperature is 10–18°C (50–65°F).[19,36,74] A WBGT meter can be used to evaluate the risk of hypothermia (see "Postscript: Measurement of Environmental Stress," page 310). Cold wind increases heat loss in proportion to wind speed; i.e., windchill factor. The relative degree of danger can be assessed (see figure 3.2 on page 75).[55] Wind speed can be estimated; if you feel the wind in your face the speed is at least 16 km · h^{-1} (kph) [10 mi · h^{-1} (mph)]; if small tree branches move or if snow and dust are raised, approximately 32 kph (20 mph); if large tree branches move, 48 kph (30 mph); if an entire tree bends, about 64 kph (40 mph)[9].

To reduce heat loss, runners should protect themselves from moisture, wind, and cold air by wearing several layers of light, loose clothing that insulate the skin with trapped air.[17] An outer garment that is windproof, allows moisture to escape, and provides rain protection is useful. Lightweight nylon parkas may not offer thermal insulation but offer significant protection against severe windchill, especially if a hood is provided. Wool and polyester fabrics retain some protective value when wet; cotton and goose down do not.[10] Areas of the body that lose large amounts of heat (head, neck, legs, hands) should be covered.[17]

Mild [34–36°C (93–97°F)] or moderate [30–34°C (86–93°F)] hypothermia should be treated before it progresses. Wet clothing should be replaced with dry material (sweatsuit, blanket) that is insulated from the ground and wind. Warm fluids should be consumed if patients are conscious, able to talk, and thinking clearly. Patients with moderate and severe [<30°C (86°F)] hypothermia should be insulated in a blanket and evacuated to a hospital immediately.[19,58] Although severe hypothermia should be treated in the field,[27] it is widely recognized that life-threatening ventricular fibrillation is common in this state and may be initiated by physical manipulation, chest compression, or intubation.[10,27,58,93] However, with conclusive evidence of cardiac standstill and breathlessness, emergency procedures (i.e., Basic Life Support, Advanced Cardiac Life Support) should be initiated. Life-support procedures[27] and commonly observed laboratory (i.e., electrolyte, acid-base) values [10,58] have been described by others.

4. Frostbite

Frostbite involves crystallization of fluids in the skin or subcutaneous tissue after exposure to subfreezing temperatures [<–0.6°C (31°F)]. With low skin temperature and dehydration, cutaneous blood vessels constrict and circulation is attenuated because the viscosity of blood increases.[55] Frostbite may occur within seconds or hours of exposure, depending upon air temperature, wind speed, and body insulation. Frostbitten skin can appear white, yellow-white, or purple, and is hard, cold, and insensitive to touch.[55] Rewarming results in intense pain, skin reddening, and swelling. Blister formation is common and loss of extremities (fingers, toes, ears, hands, feet) is possible.[36,55] The degree of tissue damage depends on duration and severity of the freezing and effectiveness of treatment.

No data have been published regarding the incidence of frostbite among athletes during training or competition. Since winter running races are rarely postponed when environmental conditions are harsh, and frostbite is the most common cold injury in military settings,[35] it is imperative that runners be aware of the dangers. Cross-country ski races are postponed if the temperature at the coldest point of the course is less than –20°C (–4°F), due to the severe windchill generated at race pace.

Runners risk frozen flesh within minutes if the air temperature and wind speed combine to present a severe windchill. Because runners prefer to have unrestricted movement during races, and because they know that exercise results in body heating, they may not wear sufficient clothing. Runners can avoid frostbite and hypothermia in cold and windy conditions by protecting themselves by dressing adequately: wet skin or clothing will increase the risk of frostbite.[21,29]

When tissue freezes [skin temperature –2 to 0°C, (28–32°F)], water is drawn out of the cells and ice crystals cause mechanical destruction of skin and subcutaneous tissue.[36] However, initial ice crystal formation is not as damaging to tissues as partial rethawing and refreezing.[40] Therefore, the decision to treat severe frostbite in the field (versus transport to a hospital) should consider the possibility of refreezing. If there is no likelihood of refreezing, the tissue should be rapidly rewarmed[36,40] in circulating warm water (40–43.3°C, 104–110°F), insulated, and the patient transported to a medical facility. Research on animals suggests that topical aloe vera and systemic ibuprofen may reduce tissue damage and speed rehabilitation in humans.[9] Other aspects of hospital treatment protocols are detailed elsewhere.[9,36,40]

RACE ORGANIZATION

The following suggestions constitute the ideal race medical team. They are offered for consideration, but are not intended as absolute requirements. Staff and equipment needs are unique to each race and may be revised after 1–2 yr, in light of the distinctive features of each race. Depending on the weather conditions, 2–12% of all entrants will typically enter a medical aid station.[1,45,50,74]

1. Medical Personnel

a. Provide medical assistance if the race is 10 km (6.2 mi) or longer.

b. Provide the following medical personnel per 1000 runners: 1–2 physicians, 4–6 podiatrists, 1–4 emergency medical technicians, 2–4 nurses, 3–6 physical therapists, 3–6 athletic trainers, and 1–3 assistants. Approximately 75% of these personnel should be stationed at the finish area. Recruit one nurse (per 1000 runners) trained in IV therapy.

c. Recruit emergency personnel from existing organizations (police, fire-rescue, emergency medical service).

d. One physician and 10–15 medical assistants serve as the triage team in the finish chute. Runners unable to walk are transported to the medical tent via wheelchair, litter, or two-person carry.

e. Consider one or two physicians and two to four nurses trained in the rehabilitative medical care of wheelchair athletes.

f. Medical volunteers should attend a briefing prior to the event to meet their supervisor and receive identification tags, weather forecast, instructions, and schedules. Supervisors from the following groups should be introduced: medical director; podiatry, nursing, physical therapy, athletic training, medical records, triage, wheelchair athlete care, and medical security (optional: chiropractic, massage therapy). Medical volunteers should be distinguished from other race volunteers; luminous/distinctive vests, coats, or hats work well.

2. Medical Aid Stations

a. Provide a primary medical aid station (250–1500 ft^2 (23–139 m^2) for each 1000 runners; see table B.1, page 300) at the finish area, with no public access. Place security guards at all entrances with instructions regarding who can enter.

b. Position secondary medical aid stations along the route at 2- to 3-km (1.2- to 1.9-mi) intervals for races over 10 km, and at the halfway point for shorter races (see table B.1). Some race directors have successfully secured equipment and medical volunteers from military reserve or national guard medical units, the American Red Cross, and the National Ski Patrol.

c. Station one ambulance per 3000 runners at the finish area and one or more mobile emergency response vehicles on the course. Staff each vehicle with a nurse and radio person or cellular telephone. Stock each vehicle with a medical kit, automatic defibrillator, IV apparatus, blankets, towels, crushed ice, blood pressure cuffs, rehydration fluid, and cups.

d. Signs should be posted at the starting line and at each medical aid station to announce the risk of heat illness or cold injury (see Appendix A).

e. A medical record card should be completed for each runner who receives treatment.[1,74] This card provides details that can be used to plan the medical coverage of future events.

f. Provide personal protective equipment (gloves, gowns, face shields, eye protection) and hand washing facilities.

g. Provide portable latrines and containers for patients with vomiting and diarrhea.

h. Initial medical assessment must include rectal (not oral, aural, or axillary temperature),[8,76] central nervous system function, and cardiovascular function. Rehydration and cooling or warming are the cornerstones of treatment.[32,41,42,50,74,94]

Table B.1
Medical Aid Stations: Suggested Equipment and Supplies Per 1000 Runners[a]

Item	Secondary aid station[b]	Primary aid station[c]
Stretchers (at 10 km and beyond)	2-5	4-10
Cots	10	30
Wheelchairs	0	1
Wool blankets (at 10 km and beyond)	6-10	12-20
Bath towels	5-10	10-20
High and low temperature rectal thermometers (37-43°C; 99-110°F) and (22-37°C; 72-99°F)[d]	5	10
Elastic bandages (2, 4, and 6 in.)	3 each	6 each
Gauze pads (4 × 4 in.)	1/2 case	1 case
Adhesive tape (1.5 in.)	1/2 case	1 case
Skin disinfectant	1 L	2 L
Surgical soap	1/2 case	1 case
Band-aids	110	220
Moleskin	1/2 case	1 case
Petroleum jelly, ointments	1/2 case	1 case
Disposable latex gloves	80 pairs	175 pairs
Stethoscopes	1	2
Blood pressure cuffs	1	2
Intravenous (IV) stations[d]	1	2
IV fluid (D5: 1/2 NS or D5:NS; 0.5 or 1 L)[d]	15[e]	30[e]
Sharps and biohazard disposal containers[d]	1	2
Alcohol wipes	50	100
Small instrument kits	1	1
Athletic trainer's kit	1	1

Item	Secondary aid station[b]	Primary aid station[c]
Podiatrist's kit	1-2	2-4
Inflatable arm and leg splints	2 each	2 each
Tables for medical supplies	1	2
Hose with spray nozzle, running water[e]	1	2
Wading pool for water immersion[e]	1	2
Fans for cooling	1	2-4
Oxygen tanks with regulators and masks	0	2
Crushed ice in plasic bags	7 kg	14 kg
Rehydration fluids	50 L[e]	100 L[e]
Cups (0.3 L, 10 oz)	1250	2250
Eye drops	1	1
Urine dipsticks[d]	10	20
Glucose blood monitoring kits[d]	1	2
Inhalation therapy for asthmatics[d]	1	1
EMS ambulance or ACLS station	1	1
Injectable drugs[d]		
Oral drugs[d]		

[a] Revised from Adner, M.M., J.J. Scarlet, J. Casey, W. Robison, and B.H. Jones. The Boston Marathon medical care team: ten years of experience. *Physician Sportsmed.* 16:99-106, 1988; Bodishbaugh, R.G. Boston marathoners get red carpet treatment in the medical tent. *Physician Sportmed.* 16:139-143, 1988; and Noble, H.B. and D. Bachman. Medical aspects of distance race planning. *Physician Sportmed.* 7:78-84, 1979.

[b] Increase supplies and equipment if the race course is out and back.

[c] At finish area.

[d] Supervised by a physician.

[e] Depends on environmental conditions.

3. Universal Precautions

All medical personnel may encounter blood-born pathogens or other potentially infectious materials, and should observe the following precautions:[53,63]

a. Receive immunization against the hepatitis B virus prior to the event.

b. Recognize that blood and infectious body fluids may be encountered from needle sticks, cuts, abrasions, blisters, and clothing.

c. Reduce the likelihood of exposure by planning tasks carefully (e.g., prohibiting recapping of needles by a two-handed technique, minimizing splashing and spraying).

d. Wear personal protective equipment such as gloves, gowns, face shields, and eye protection. Remove this equipment and dispose/decontaminate it prior to leaving the work area.

e. Wash hands after removing gloves or other personal protective equipment.

f. Dispose of protective coverings, needles, scalpels, and other sharp objects in approved, labeled biohazard containers.

g. Do not eat, drink, smoke, handle contact lenses, or apply cosmetics/lip balm in the medical treatment area.

h. Decontaminate work surfaces, bins, pails, and cans [1/10 solution of household bleach (sodium hypochlorite) in water] after completion of procedures.

4. Fluid Stations

a. At the start and finish areas provide 0.34–0.45 L (12–16 oz) of fluid per runner. At each fluid station on the race course (2–3 km apart), provide 0.28–0.34 L (10–12 oz) of fluid per runner. Provide both water and a carbohydrate-electrolyte beverage in equal volumes.

b. In cool or cold weather [10°C (50°F)], an equivalent amount of warm fluid should be available.

c. Number of cups (0.3 L, 10 oz) per fluid station on the course = number of entrants + 25% additional for spillage and double use. Double this total if the course is out and back.

d. Number of cups at start and finish area = (2 · number of entrants) + 25% additional.

e. Cups should be filled prior to the race and placed on tables to allow easy access. Runners drink larger volumes if volunteers hand them cups filled with fluid.

5. Communications/Surveillance

a. Provide two-way radio or telephone communication between the medical director, medical aid stations, mobile vans, and pick-up vehicles.

b. Arrange for radio-equipped vehicles to drive the race course (ahead and behind participants) and provide communication with the director and his/her staff. These vehicles should be stationed at regular intervals along the course to search the course for competitors who require emergency care and encourage compromised runners to stop.

c. Place radio-equipped observers along the course.

d. Notify local hospitals, police, and fire-rescue departments of the time of the event, number of participants, location of aid stations, extent of medical coverage, and the race course.

e. Use the emergency response system (telephone number 911) in urban areas.

6. Instructions to Runners

a. Advise each race participant to print name, address, telephone number, and medical problems on the back of the race number (pinned to the body). This permits emergency personnel to quickly identify unconscious runners. Inform emergency personnel that this information exists.

b. Inform race participants of potential medical problems at pre-race conferences and at the starting line. Signed registration forms should clearly state the types of heat or cold injuries that may arise from participation in this event.

c. Provide pre-event recommendations regarding training, fluid consumption, clothing selection, self-care, heat acclimatization, and signs or symptoms of heat/cold illness.[88]

d. The race director should announce the following information to all participants by loudspeaker immediately prior to the race:

- Current and predicted maximum (or minimum) temperature, humidity, wind speed, and cloud cover;
- The WBGT category and the risks for hyperthermia or hypothermia (see Appendix A);
- Location of aid stations, types of assistance, and fluid availability;
- Signs and symptoms of heat or cold illness;
- Recommended clothing;
- The need for fluid replacement before, during, and after the race;

- The policy of race monitors to stop runners who are ill;
- A request that runners seek help for impaired competitors who appear ill, who are not coherent, who run in the wrong direction, or who exhibit upper-body swaying and poor competitive posture;
- A warning to novice runners entering their first race that they should run at a comfortable pace and run with a partner;
- Warnings to runners who are taking medications or have chronic illnesses (asthma, hypertension, diabetes, cardiovascular problems).

ACKNOWLEDGMENT

This position stand replaces the 1987 ACSM position paper, "The Prevention of Thermal Injuries During Distance Running."

This pronouncement was reviewed for the American College of Sports Medicine by members-at-large, the Pronouncements Committee, and by: Arthur E. Crago, MD; Stafford W. Dobbin, MD; Mary L. O'Toole, PhD, FACSM; LTC Katy L. Reynolds, MD; and John W. Robertson, MD, FACSM.

REFERENCES

1. Adner, M. M., J. J. Scarlet, J. Casey, W. Robison, & B. H. Jones. 1988. The Boston Marathon medical care team: Ten years of experience. *Physician Sportsmed. 16*: 99-106.

2. Adolph, E. F. 1947. *Physiology of man in the desert*, 5-43. New York: Interscience.

3. American College of Sports Medicine. 1996. Position stand: Exercise and fluid replacement. *Med. Sci. Sports Exerc. 28*: i-vii.

4. American College of Sports Medicine. 1987. Position stand: The prevention of thermal injuries during distance running. *Med. Sci. Sports Exerc. 19*: 529-533.

5. Armstrong, L. E., J. P. de Luca, & R. W. Hubbard. 1990. Time course of recovery and heat acclimation ability of prior exertional heatstroke patients. *Med. Sci. Sports Exerc. 22*: 36-48.

6. Armstrong, L. E. & C. M. Maresh. 1991. The induction and decay of heat acclimatization in trained athletes. *Sports Med. 12*: 302-312.

7. Armstrong, L. E. & C. M. Maresh. 1993. The exertional heat illnesses: A risk of athletic participation. *Med. Exerc. Nutr. Health 2*: 125-134.

8. Armstrong, L. E., C. M. Maresh, A. E. Crago, R. Adams, & W. O. Roberts. 1994. Interpretation of aural temperatures during exercise, hyperthermia, and cooling therapy. *Med. Exerc. Nutr. Health 3*: 9–16.

9. Bangs, C. C., J. A. Boswick, M. P. Hamlet, D. S. Sumner, R. C. A. Weatherly-white, & W. J. Mills. 1977. When your patient suffers frostbite. *Patient Care 12*: 132–157.

10. Bangs, C., M. P. Hamlet, & W. J. Mills. 1977. Help for the victim of hypothermia. *Patient Care 12*: 46–50.

11. Bar-Or, O. 1980. Climate and the exercising child—a review. *Int. J. Sports Med. 1*: 53–65.

12. Bar-Or, O. 1989.Temperature regulation during exercise in children and adults. In *Perspectives in exercise science and sports medicine, Vol. 2, Youth exercise and sport*, edited by C. V. Gisolfi & D. R. Lamb 335–367. Indianapolis: Benchmark Press.

13. Bar-Or, O., H. M. Lundegren, & E. R. Buskirk. 1969. Heat tolerance of exercising lean and obese women. *J. Appl. Physiol. 26*: 403–409.

14. Bassett, D. R., F. J. Nagle, S. Mookerjee, et al. 1988. Thermoregulatory responses to skin wetting during prolonged treadmill running. *Med. Sci. Sports Exerc. 19*: 28–32.

15. Bodishbaugh, R. G. 1988. Boston marathoners get red carpet treatment in the medical tent. *Physician Sportsmed. 16*: 139–143.

16. Brengelmann, G. L. 1987. Dilemma of body temperature measurement. In *Man in stressful environments: Thermal and work physiology*, edited by K. Shiraki & M. K. Yousef, 5–22. Springfield, IL: Charles C Thomas.

17. Buckley, R. L. & R. Hostetler. 1990. The physiologic impact and treatment of hypothermia. *Med. Times 118*: 38–44.

18. Burns, J. P. 1988. Liability pertaining to endurance athletic events. In *Medical coverage of endurance athletic events*, edited by R. H. Laird, 62–68. Columbus, OH: Ross Laboratories.

19. Burr, L. 1983. Accidental hypothermia: Always a danger. *Patient Care 17*: 116–153.

20. Buskirk, E. R., P. F. Iampietro, & D. E. Bass. 1958. Work performance after dehydration: Effects of physical conditioning and heat acclimatization. *J. Appl. Physiol. 12*: 189–194.

21. Casey, M. J., C. Foster, & E. G. Hixon (eds.). 1990. *Winter sports medicine*, 1–450. Philadelphia: F. A. Davis Co.

22. Clowes, G. H. A., Jr. & T. F. O'Donnell, Jr. 1974. Heat stroke. *N. Engl. J. Med. 291*: 564–567.

23. Costill, D. L., R. Cote, E. Miller, T. Miller, & S. Wynder. 1970. Water and electrolyte replacement during days of work in the heat. *Aviat. Space Environ. Med. 46*: 795–800.

24. Costill, D. L., W. F. Kammer, & A. Fisher. 1970. Fluid ingestion during distance running. *Arch. Environ. Health 21*: 520–525.

25. Costrini, A. 1990. Emergency treatment of exertional heatstroke and comparison of whole body cooling techniques. *Med. Sci. Sports Exerc. 22*: 15–18.

26. Costrini, A. M., H. A. Pitt, A. B. Gustafson, & D. E. Uddin. 1979. Cardiovascular and metabolic manifestations of heatstroke and severe heat exhaustion. *Am. J. Med. 66*: 296–302.

27. Cummins, R. O. (ed.). 1994. *Textbook of advanced cardiac life support,* 10/10–10/12. Dallas: American Heart Association.

28. Department of the Army. 1980. *Prevention, treatment and control of heat injury.* Washington, DC: Department of the Army, Technical Bulletin No. TB MED 507: 1–21.

29. Dobbin, S. W. 1986. Providing medical services for fun runs and marathons in North America. In *Sports medicine for the mature athlete,* edited by J. R. Sutton & R. M. Brock, 193–203. Indianapolis: Benchmark Press.

30. England, A. C. III, D. W. Fraser, A. W. Hightower, et al. 1982. Preventing severe heat injury in runners: Suggestions from the 1979 Peachtree Road Race experience. *Ann. Intern. Med 97*: 196–201.

31. Epstein, Y. Heat intolerance: Predisposing factor or residual injury? 1990. *Med. Sci. Sports Exerc. 22*: 29–35.

32. Epstein, Y., E. Sohar, & Y. Shapiro. Exertional heatstroke: A preventable condition. 1995. *Isr. J. Med. Sci. 31*: 454–462.

33. Gisolfi, C. V. & J. R. Cohen. 1979. Relationships among training, heat acclimation and heat tolerance in men and women: The controversy revisited. *Med. Sci. Sports 11*: 56–59.

34. Gisolfi, C. V. & J. R. Copping. 1974. Thermal effects of prolonged treadmill exercise in the heat. *Med. Sci. Sports 6*: 108–113.

35. Hamlet, M. 1987. An overview of medically related problems in the cold environment. *Mil. Med. 152*: 393–396.

36. Hamlet, M. P. 1988. Human cold injuries. In *Human performance physiology and environmental medicine at terrestrial extremes,* edited by K B. Pandolf, M. N. Sawka, & R. R. Gonzalez, 435–466. Indianapolis: Benchmark Press.

37. Hanson, P. G. & S. W. Zimmerman. Exertional heatstroke in novice runners. 1979. *J.A.M.A. 242*: 154–157.

38. Hart, L. E., B. P. Egier, A. G. Shimizu, P. J. Tandan, & J. R. Sutton. 1980. Exertional heat stroke: The runner's nemesis. *Can. Med. Assoc. J. 122*: 1144–1150.

39. Haymes, E. M., R. J. McCormick, & E. R. Buskirk. 1975. Heat tolerance of exercising lean and obese prepubertal boys. *J. Appl. Physiol. 39*: 457–461.

40. Heggers, J. P., L. G. Phillips, R. L. McCauley, & M. C. Robson. 1990. Frostbite: Experimental and clinical evaluations of treatment. *J. Wilderness Med. 1*: 27–32.

41. Hubbard, R. W. & L. E. Armstrong. 1988. The heat illnesses: Biochemical, ultrastructural, and fluid-electrolyte considerations. In *Human performance physiology and environmental medicine at terrestrial extremes,* edited by K.

B. Pandolf, M. N. Sawka, & R. R. Gonzalez, 305–359. Indianapolis: Benchmark Press.

42. Hubbard, R. W. & L. E. Armstrong. 1989. Hyperthermia: New thoughts on an old problem. *Physician Sportsmed. 17*: 97–113.

43. Hughson, R. L., H. J. Green, M. E. Houston, J. A. Thomson, D. R. Maclean, & J. R. Sutton. 1980. Heat injuries in Canadian mass participation runs. *Can. Med. Assoc. J. 122*: 1141–1144.

44. Hughson, R. L., L. A. Standi, & J. M. Mackie. 1983. Monitoring road racing in the heat. *Physician Sportsmed. 11*: 94–105.

45. Jones, B. H., P. B. Rock, L. S. Smith, et al. 1985. Medical complaints after a marathon run in cool weather. *Physician Sportsmed. 13:* 103–110.

46. Keren, G., Y. Epstein, & A. Magazanik. 1981. Temporary heat intolerance in a heatstroke patient. *Aviat. Space Environ. Med. 52*: 16–117.

47. Khogali, M. 1983. The Makkah body cooling unit. In *Heat stroke and temperature regulation,* edited by M. Khogali & J. R. S. Hales, 139–148. New York: Academic Press.

48. Khogali, M. & J. S. Weiner. 1980. Heat stroke: Report on 18 cases. *Lancet 2*: 276–278.

49. Kielblock, A. J. 1987. Strategies for the prevention of heat disorders with particular reference to body cooling procedures. In *Heat stress,* edited by J. R. S. Hales & D. A. B. Richards, 489–497. Amsterdam: Elsevier.

50. Kleiner, D. M. & S. E. Glickman. 1994. Medical considerations and planning for short distance road races. *J. Athl. Train. 29:*145–151.

51. Knochel, J. P. 1974. Environmental heat illness: An eclectic review. *Arch. Intern. Med. 133*: 841–864.

52. Maron, M. B., J. A. Wagner, & S. M. Horvath. 1977. Thermoregulatory responses during competitive distance running. *J. Appl. Physiol. 42*: 909–914.

53. Massachusetts Medical Society. 1988. Update: Universal precautions for prevention of transmission of Human Immunodeficiency Virus, Hepatitis B Virus, and other bloodborne pathogens in healthcare settings. *MMWR 37*: 377–388.

54. Maughan, R. J., I. M. Light, P. H. Whiting, & J. D. B. Miller. 1982. Hypothermia, hyperkalemia, and marathon running. *Lancet 11*: 1336.

55. Milesko-Pytel, D. 1983. Helping the frostbitten patient. *Patient Care 17*: 90–115.

56. Nadel, E. R., C. B. Wenger, M. F. Roberts, J. A. J. Stolwijk, & E. Cafarelli. 1977. Physiological defenses against hyperthermia of exercise. *Ann. N. Y. Acad. Sci. 301*: 98–109.

57. Nash, H. L. 1985. Treating thermal injury: Disagreement heats up. *Physician Sportsmed. 13*: 134–144.

58. Nelson, R. N. 1985. Accidental hypothermia. In *Environmental emergencies,* edited by R. N. Nelson, D. A. Rund, & M. D. Keller, 1–40. Philadelphia: W. B. Saunders Co.

59. Noakes, T. D. 1986. Body cooling as a method for reducing hyperthermia. *S. Afr. Med. J. 70*: 373–374.

60. Noakes, T. D., N. Goodwin, B. L. Rayner, T. Branken, & R. K. N. Taylor. 1985. Water intoxication: A possible complication during endurance exercise. *Med. Sci. Sports Exerc. 17*: 370–375.

61. Noble, H. B. & D. Bachman. 1979. Medical aspects of distance race planning. *Physician Sportsmed. 7*: 78–84.

62. O'Donnell, T. J., Jr. 1975. Acute heatstroke. Epidemiologic, biochemical, renal and coagulation studies. *J.A.M.A. 234*: 824–828.

63. Occupational Safety and Health Administration. 1991. Occupational exposure to bloodborne pathogens; final rule. *Fed. Register 56*: 64175–64182.

64. Pandolf, K. B., R. L. Burse, & R. F. Goldman. 1977. Role of physical fitness in heat acclimatization, decay and reinduction. *Ergonomics 20*: 399–408.

65. Payen, J., L. Bourdon, H. Reutenauer, et al. 1992. Exertional heatstroke and muscle metabolism: An in vivo ^{31}P-MRS study. *Med. Sci. Sports Exerc. 24*: 420–425.

66. Pearlmutter, E. M. 1986. The Pittsburgh marathon: "Playing weather roulette." *Physician Sportsmed. 14*: 132–138.

67. Piwonka, R. W., S. Robinson, V. L. Gay, & R. S. Manalis. 1965. Preacclimatization of men to heat by training. *J. Appl. Physiol. 20*: 379–384.

68. Pugh, L. G. C. E. 1967. Cold stress and muscular exercise, with special reference to accidental hypothermia. *Br. Med. J. 2*: 333–337.

69. Pugh, L. G. C. E., J. L. Corbett, & R. H. Johnson. 1967. Rectal temperatures, weight losses and sweat rates in marathon running. *J. Appl. Physiol. 23*: 347–352.

70. Richards, R. & D. Richards. 1984. Exertion-induced heat exhaustion and other medical aspects of the city-to-surf fun runs, 1978-1984. *Med. J. Aust. 141*: 799–805.

71. Richards, R., D. Richards, P. J. Schofield, V. Ross, & J. R. Sutton. 1979. Reducing the hazards in Sydney's *The Sun* city-to-surf runs, 1971 to 1979. *Med. J. Aust. 2*: 453–457.

72. Richards, D., R. Richards, P. J. Schofield, V. Ross, & J. R. Sutton. 1979. Management of heat exhaustion in Sydney's *The Sun* city-to-surf fun runners. *Med. J. Aust. 2*: 457–461.

73. Richards, R., D. Richards, P. J. Schofield, V. Ross, & J. R. Sutton. 1979. Organization of *The Sun* city-to-surf fun run, Sydney. *Med. J. Aust. 2*: 470–474.

74. Roberts, W. O. 1989. Exercise-associated collapse in endurance events: A classification system. *Physician Sportsmed. 17*: 49–55.

75. Roberts, W. O. 1992. Managing heatstroke: On-site cooling. *Physician Sportsmed. 20*: 17–28.

76. Roberts, W. O. 1994. Assessing core temperature in collapsed athletes. *Physician Sportsmed. 22*: 49–55.

77. Robinson, S., S. L. Wiley, L. G. Boudurant, & S. Mamlin, Jr. 1976. Temperature regulation of men following heatstroke. *Isr. J. Med. Sci. 12*: 786–795.

78. Roos, R. 1987. Medical coverage of endurance events. *Physician Sportsmed. 15*: 140–147.

79. Royburt, M., Y. Epstein, Z. Solomon, & J. Shemer. 1993. Long term psychological and physiological effects of heat stroke. *Physiol. Behav. 54*: 265–267.

80. Sandell, R. C., M. D. Pascoe, & T. D. Noakes. 1988. Factors associated with collapse during and after ultramarathon footraces: A preliminary study. *Physician Sportsmed. 16*: 86–94.

81. Sawka, M. N. & K. B. Pandolf. 1990. Effects of body water loss on physiological function and exercise performance. In *Perspectives in exercise science and sports medicine. Vol. 3. Fluid homeostasis during exercise,* edited by C. V. Gislofi & D. R. Lamb, 1–38. Carmel, IN: Benchmark Press.

82. Shapiro, Y., A. Magazanik, R. Udassin, G. Ben-Baruch, E. Shvartz, & Y. Shoenfeld. 1979. Heat tolerance in former heatstroke patients. *Ann. Intern. Med. 90*: 913–916.

83. Shibolet, S., R. Coll, T. Gilat, & E. Sohar. 1967. Heatstroke: Its clinical picture and mechanism in 36 cases. *Q. J. Med. 36*: 525–547.

84. Shibolet, S., M. C. Lancaster, & Y. Danon. 1976. Heat stroke: A review. *Aviat. Space Environ. Med. 47*: 280–301.

85. Shvartz, E., Y. Shapiro, A. Magazanik, et al. 1977. Heat acclimation, physical fitness, and responses to exercise in temperate and hot environments. *J. Appl. Physiol. 43*: 678–683.

86. Sutton, J. R. 1984. Heat illness. In *Sports medicine,* edited by R. H. Strauss, 307–322. Philadelphia: W. B. Saunders.

87. Sutton, J. R. 1986. Thermal problems in the masters athlete. In *Sports medicine for the mature athlete,* edited by J. R. Sutton & R. M, Brock. 125–132. Indianapolis: Benchmark Press.

88. Sutton J. R. 1990. Clinical implications of fluid imbalance. In *Perspectives in exercise science and sports medicine. Vol. 3. Fluid homeostasis during exercise,* edited by C. V. Gisolfi & D. R. Lamb, 425–455. Carmel, IN: Benchmark Press.

89. Sutton, J. R. & O. Bar-Or. 1980. Thermal illness in fun running. *Am. Heart J. 100*: 778–781.

90. Sutton, J, R., M. J. Coleman, A. P. Millar, L. Lazarus, & P. Russo. 1972. The medical problems of mass participation in athletic competition. The "City-to-Surf" race. *Med. J. Aust. 2*: 127–133.

91. Thompson, P. D., M. P. Stern, P. Williams, K. Duncan, W. L. Haskell, & P. D. Wood. 1979. Death during jogging or running. A study of 18 cases. *J.A.M.A. 242*: 1265–1267.

92. Weinberg, S. 1988. The Chicago Bud Light Triathlon. In *Medical coverage of endurance athletic events,* edited by R. H. Laird, 74–79. Columbus, OH: Ross Laboratories.

93. White, J. D. & F. S. Southwick. 1980. Cardiac arrest in hypothermia. *J.A.M.A.* *244*: 2262.

94. Winslow, C. E. A., L. P. Herrington, & A. P. Gagge. 1937. Physiological reactions of the human body to various atmospheric humidities. *Am. J. Physiol.* *120*: 288–299.

95. Wyndham, C. H. & N. B. Strydom. 1969. The danger of inadequate water intake during marathon running. *S. Afr. Med. J. 43*: 893–896.

96. Yaglou, C. P. & D. Minard. 1957. Control of heat casualties at military training centers. *Arch. Ind. Health 16*: 302–305.

97. Yarbrough, B. E. & R. W. Hubbard. 1989. Heat-related illnesses. In *Management of wilderness and environmental emergencies,* 2nd ed., edited by P. S. Auerbach & E. C. Geehr, 119–143. St. Louis: C. V. Mosby Co.

POSTSCIPT: MEASUREMENT OF ENVIRONMENTAL STRESS

Ambient temperature is only one component of environmental heat or cold stress; others are humidity, wind speed, and radiant heat. The most widely used heat stress index is the wet bulb globe temperature (WBGT) index:[96]

$$WBGT = (0.7\, T_{wb}) + (0.2\, T_{g}) + (0.1 T_{db})$$

where T_{wb} is the wet bulb temperature, T_{g} is the black-globe temperature, and T_{db} is the shaded dry bulb temperature.[28] T_{db} refers to air temperature measured with a standard dry bulb thermometer not in direct sunlight. T_{wb} is measured with a water-saturated cloth wick over a dry bulb thermometer (not immersed in water). T_{g} is measured by inserting a dry bulb thermometer into a standard black metal globe. Both T_{wb} and T_{g} are measured in direct sunlight.

A portable monitor that gives the WBGT index in degrees Celsius or degrees Fahrenheit has proven useful during races and in military training.[28,44,87,96] The measurement of air temperature alone is inadequate. The importance of humidity in total heat stress can be readily appreciated because T_{wb} accounts for 70% of the index whereas T_{db} accounts for only 10%.

The risk of heat illness (while wearing shorts, socks, shoes, and a T-shirt) due to environmental stress should be communicated to runners in four categories (see figure 2.9 on page 47):

- Very high risk: WBGT above 28°C (82°F);
- High risk: WBGT 23–28°C (73–82°F);

- Moderate risk: WBGT 18–23°C (65–73°F);
- Low risk: WBGT below 18°C (65°F).

Large signs should be displayed, at the start of the race and at key points along the race course, to describe the risk of heat exhaustion and heatstroke (figure 2.9). When the WBGT index is above 28°C (82°F), the risk of heat exhaustion or heatstroke is very high; it is recommended that the race be postponed until less stressful conditions prevail, rescheduled, or canceled. High risk [WBGT index = 23–28°C (73–82°F)] indicates that runners should be aware that heat exhaustion or heatstroke may be experienced by any participant; anyone who is particularly sensitive to heat or humidity probably should not run. Moderate risk [WBGT index = 18–23°C (65–73°F)] reminds runners that heat and humidity will increase during the course of the race if conducted during the morning or early afternoon. Low risk [WBGT index below 18°C (65°F)] does not guarantee that heat exhaustion (even heatstroke, see ref. 5,32) will not occur; it only indicates that the risk is low.

The risk of hypothermia (while wearing shorts, socks, shoes, and a T-shirt) also should be communicated to runners. A WBGT index below 10°C (50°F) indicates that hypothermia may occur in slow runners who run long distances, especially in wet and windy conditions. Core body temperatures as low as 92°F have been observed in 65°F conditions.[74]

GLOSSARY

acclimation—**Acclimatization** in an artificial environment (e.g., an environmental chamber).

acclimatization—A complex of adaptive responses that demonstrate improved homeostatic balance in multiple organs; usually requires 10-14 days for responses to develop adequately; your body can acclimatize (to varying degrees) to hot, cold, high-altitude, underwater, and air-polluted environments.

accommodation—An immediate physiological change in the sensitivity of a cell or tissue to change(s) in the external environment.

acute mountain sickness (AMS)—A complex of symptoms that occur usually in unacclimatized sea-level residents who ascend rapidly to high altitude; AMS is common over 3000 m; moderate to severe cases involve fluid retention, increased intracranial pressure, and lung swelling (i.e., edema); severe AMS involves fatigue, headache, apathy, and irritability.

adaptive responses (adaptations)—Physiological changes that minimize bodily strain that may be short-term, intermediate, or long-term.

adrenal cortex—The outer 80% of each adrenal gland; secretes the hormones cortisol (which alters metabolism of protein and carbohydrates in response to stress), aldostereone (which maintains sodium balance and extracellular fluid volume), and precursors to testosterone and estradiol (sex hormones).

adrenal medulla—The inner core of each adrenal gland; secretes the hormones epinephrine and norepinephrine in response to stimulation of the sympathetic nervous system; important in the "fight or flight" reaction to stress or threat.

aerobic metabolism—With oxygen; usually refers to generation of ATP within mitochondria.

aerosol—A suspension of ultramicroscopic solid or liquid particles in air or another gas; examples include smoke, fog, or mist; the majority of scientific studies have investigated either sulfate, sulfuric acid (H_2SO_4), or nitrate aerosols.

air embolism—A type of barotrauma in which the gases in the lung expand during ascent. It involves the expansion of lung gases during ascent, causing alveoli rupture and forcing air bubbles into the blood. May cause heart attack or stroke.

air pressure—The measurable weight of the vertical column of atmospheric gases (e.g., air) above your body.

alternobaric vertigo (ABV)—A form of barotrauma that usually occurs during ascent and is due to the sudden development of unequal middle-ear pressure, causing pronounced vertigo.

altitude acclimatization—A complex of adaptions to hypoxia at high altitude that enhances performance and increases the chances of survival.

alveolar rupture—Tearing of alveoli in the lungs.

ama—A widely studied group of breath-hold divers located in Japan and Korea.

anaerobic metabolism—Without oxygen; usually refers to anaerobic glycolysis, the metabolic pathway that generates ATP by converting glucose to pyruvic acid; found in cells, outside the mitochondria.

ATP—Adenosine triphosphate; a molecule that serves as "energy currency"; ATP is used as an energy source for muscle contraction and most other functions of cells.

AVP—Arginine vasopressin; a circulating hormone that reduces water loss at the kidneys; also known as antidiuretic hormone (ADH).

barotrauma—Hyperbaric damage to organs and tissues of the body, experienced during diving.

basal metabolic rate—The amount of energy that is required to sustain life when the person is at physical, emotional, and digestive rest.

behavioral thermoregulation—Conscious acts that alter heat loss, body insulation, or ambient temperature (e.g., altering layers of clothing, drinking hot or cold liquids, moving to another environment).

biological rhythm—Normal responses of human organs and systems often are cyclic (e.g., biological rhythms) and may be influenced by cues from the environment, social interactions, and exercise; these rhythms can be desynchronized by travel across time zones and sleep loss.

biometeorology—The study of the effects of environmental factors on living organisms.

blood doping—Also known as "blood boosting" and "blood packing"; this technique attempts to increase the oxygen-carrying capacity of blood by increasing the hemoglobin content, thereby enhancing endurance performance; two methods exist: the first involves removing red blood cells (rbc) from an athlete and reinfusing them at a later date; the second method involves the use of an artificial hormone (see **erythropoietin**).

bone necrosis—An occupational health hazard that occurs among commercial divers over many months and years of hyperbaric exposure; segments of bone die at a slow rate and may not produce symptoms for months or years; this disease occurs most often at the shoulder, hip, and knee joints.

breath-hold diving—Diving underwater without use of a supplemental air supply. Also known as skin diving or free diving.

breath-hold diving blackout—A condition of lost consciousness in breath-hold divers caused by hypoxia.

breathing gas contamination—Contaminants in the pressurized breathing gas used by divers. As scuba gas cylinders are pressurized, the sea-level pressure of each individual gas is multiplied, and any contaminent in th air source can become dangerous to the diver.

bronchial asthma—A condition caused by bronchoconstriction of the airways, due to allergies, exercise, cold-dry air, or air pollutants.

bronchoconstriction—Bronchospasm; contraction of the smooth muscles surrounding the bronchial tubes; reduces the flow of oxygen into the lungs.

carbon monoxide (CO)—The most significant air pollutant; found in automobile exhaust; alters the ability of red blood cells to carry oxygen to skeletal muscle and other tissues; blocks the oxygen-binding sites on the molecule hemoglobin.

carboxyhemoglobin (HbCO)—Hemoglobin (Hb) combines readily with carbon monoxide (CO) to form carboxyhemoglobin (HbCO); because the affinity of Hb for CO is about 230 times greater than its affinity for oxygen (O_2), CO has great potential to alter oxygen transport in blood.

cardiorespiratory endurance—The ability of the body to sustain prolonged exercise.

chilblain—A minor form of cold injury that affects the legs, toes, hands, and ears; a severe cold-induced skin vasoconstriction causes lack of oxygen in cells and limb swelling; similar to pernio.

chronic mountain sickness—An illness that develops after months or years of residence at high altitude; may lead to disability or death.

circadian rhythm—Derived from the Latin phrase *circa diem*, which means "around a day"; a biological rhythm that lasts approximately 24 h (typical range: 20-28 h), suggesting that it is synchronized with the rotation of the earth.

climate—The average condition of the weather at a specific site, over a period of years.

CO_2 toxicity—A diving illness caused by elevated levels of carbon dioxide in human tissue.

cold habituation—A type of adaptation to cold exposure; if you immerse your hand or body in very cold water, you will notice that the pain and discomfort subside after a few minutes; this response involves habituation of feeling from pain-sensitive neurons in or near the skin.

compartment syndrome—Increased fluid pressure inside a muscle that causes swelling and pain; this increased size exceeds the expansion capacity of the fascia (i.e., the tissue enclosing the muscle).

conduction—Occurs when two solid objects are in direct contact and heat is transferred from the warmer to the cooler surface.

convection—Heat exchange that occurs between a solid medium and one that moves (e.g., a fluid such as air and water).

critical temperature—The water or air temperature at which shivering begins and metabolic rate (e.g., oxygen consumption and energy production) increases.

deacclimatization—The loss of adaptations gained during acclimatization; usually requires 21 days or more.

decompression sickness—Occurs when a diver returns to the surface too rapidly, causing oxygen and nitrogen gas pressures *within* tissues to exceed the *external* hydrostatic pressure; also known as "the bends."

DNA—Deoxyribonucleic acid; the molecule that contains all genetic/hereditary information.

edema—Swelling in which the body's tissues contain an excess of fluid.

effective dose—An expression of the effect of exercise on the total amount of a pollutant (i.e., O_3) inhaled; the multiplication product of the pollutant concentration (ppm), pulmonary ventilation (liters per minute), and exposure time (minutes).

entraining agent—An external cue that resets the duration of a biological rhythm; synonymous with the terms **synchronizer** and **zeitgeber**.

enzyme—A protein molecule that speeds chemical reactions.

epinephrine (adrenaline)—A hormone secreted by the adrenal medulla that prepares the body for the "fight or flight" response and mobilization/expenditure of energy (e.g., glucose and free fatty acid mobilization in blood); increases blood pressure, cardiac output, cell metabolism, and mental activity.

erythropoietin (EPO)—A hormone produced in the kidneys that stimulates increased red blood cell production in bone marrow; artificial EPO is used by some athletes as an illegal ergogenic aid to enhance endurance performance.

exercise intensity—The level at which muscular activity is maintained; quantified in terms of work output, velocity, force, power.

exertional heatstroke—A medical emergency involving life-threatening hyperthermia (rectal temperature exceeding 39.5-40°C, 103-104°F); treatment involves cooling the body.

external ear squeeze—**Suit squeeze** involving an ear.

face-mask squeeze—A hypobaric injury caused by a failure to equalize pressure in a diving mask during descent, this condition may result in ruptured blood vessels in the eye or, in severe cases, dislocation of the eyes from the orbit of the skull.

fine particulates—A type of air pollution that includes dust, cigarette smoke, and pollen; large particles (5-10 μm) are deposited in the nasopharyngeal region and cause inflammation, congestion, and ulceration; smaller particles stimulate bronchospasm, bronchial congestion, and bronchitis; only the smallest particles (0.5-3 μm) reach the alveoli.

frostbite—A form of cold injury caused by contact with supercold metal, liquid, or wind; the outer layer and/or inner layer of skin may be damaged; ice crystals form in tissues and damage cell membranes.

frostnip—A form of cold injury in which only the epidermis (outer skin layer) is frozen.

gastrointestinal barotrauma—A hyperbaric injury involving fullness, belching, flatulence, and abdominal pain due to bowel gas that expands during ascent.

gene—The physical unit of heredity, located in chromosomes; a sequence of molecular parts that code for the production of a specific protein segment of **DNA** or **RNA**.

genetic adaptation—Semipermanent morphological, physiological, or other changes that occur over many generations within one species that favor survival in a particular environment.

graded ascent—Climbing in stages; staying at intermediate altitudes before climbing higher.

habituation—A dampening of the normal response to a **stressor**.

heat acclimatization (HA)—Exposing oneself to exercise-heat stress gradually, on consecutive days, to stimulate adaptive responses that improve exercise performance and heat tolerance, and reduce physiological strain and the incidence of some forms of heat illness.

heat cramps—Muscular cramps occurring during exercise in a hot environment, due to either (a) loss of sodium chloride (NaCl) in sweat and urine and inadequate dietary intake of NaCl, (b) replacement of sweat losses with a large volume of dilute fluid or pure water, or (c) both; treatment involves restoring NaCl balance.

heat exhaustion—The most common form of heat illness, defined as the inability to continue exercise in a hot environment; caused by water and salt depletion subsequent to sweating.

heat syncope—Fainting after prolonged standing (common among laborers, soldiers, civilians, band members) or exercise in a hot environment; results from pooling of blood in the veins of the limbs and skin, leading to insufficient blood returning to the heart and brain; heat acclimatization reduces the incidence of heat syncope.

hemoglobin—A large protein molecule that carries oxygen from lungs to tissues; resides within red blood cells.

high-altitude cerebral edema (HACE)—Brain swelling; an extreme form of **acute mountain sickness (AMS)** that includes loss of consciousness; can be fatal if left untreated; occurs in about 1% of all individuals exposed to hypobaric hypoxia.

high-altitude pulmonary edema (HAPE)—A potentially fatal illness that occurs in unacclimatized lowlanders who visit high altitude by ascending rapidly; these individuals experience fluid accumulation in lung alveoli and subsequent clinical symptoms.

high-altitude retinal hemorrhage (HARH)—Capillary damage in the retina; experienced by virtually all mountain climbers who go above 6700 m; may be due to surges in blood pressure during exercise in a hypobaric environment.

homeostasis—The body's tendency to maintain a steady state despite external changes.

hormonal axis—A sequence of hormone influences in which one hormone triggers the action of subsequent hormones.

hormone—A chemical that is released into the bloodstream by an endocrine gland and exerts its influence at a distant target organ.

hydrostatic pressure—The pressure that water exerts on an immersed body; this is applied from all directions toward the center of the body.

hyperbaric—Pressure that is greater than air pressure at sea level (as in an underwater environment).

hyperbaric acclimatization—An adaptive change in lung capacity and sensitivity to changes in blood contents, due to repeated breath-hold dives.

hypercapnia—Elevated blood carbon dioxide levels.

hyperthermia—Body temperature elevated above 39°C (102°F).

hyperventilation—Repetitive, rapid, and deep air breathing; this response causes carbon dioxide to be expired in greater than normal quantities.

hypobaria—Pressure that is less than air pressure at sea level (as in a high-altitude environment).

hypohydration—A state of chronic, low body water lasting 4 or more hours without rehydration.

hyponatremia—A plasma sodium concentration below 130 mEq Na⁺/L; indicative of either a whole-body sodium deficit or dilution of the extracellular fluid; although rare, athletes typically encounter this illness in events lasting at least 7 hours.

hypothalamus—The portion of the brain that controls body temperature, thirst, hunger, and other processes.

hypothermia—Mild hypothermia occurs when core body temperature drops from 37°C to 32-35°C (from 98.6°F to 90-95°F); the symptoms include confusion, apathy, withdrawal, and slurred speech; death occurs at a core temperature of about 22°C (72°F).

hypoxia—Inadequate oxygen delivery to the body's tissues.

hypoxic—Oxygen content of air that is lower than that found at sea level.

immersion foot—A form of cold injury caused by prolonged exposure of wet feet to a temperature of 0.5-10°C (33-50°F); nerves and blood vessels are damaged; similar to trench foot.

inert gas—A gas that reacts poorly with other elements (i.e., nitrogen, helium, neon, argon, xenon, krypton, hydrogen).

inert gas narcosis—A euphoric state, similar to **nitrogen narcosis**, caused by breathing inert gas(es).

inflammation—A tissue's reaction to injury from heat, cold, solar (e.g., ultraviolet) radiation, microorganisms, surgery, or electricity.

insulative acclimatization—A type of adaptation to cold exposure; change(s) occur in either skin blood flow or subcutaneous fat to protect and insulate the body.

interleukin-6—A cell involved in immune function; stimulates the growth and action of other immune cells (e.g., T cells and B cells).

in-water decompression—A method of safely returning from a deep sea dive; it involves stopping at specific depths during ascent.

ion—A particle, compound, or element that carries an electrical charge (i.e., positive or negative); gas molecules may be divided into ions by ultraviolet rays, gamma rays, or X rays.

jet lag—A change in light-dark cycles and training/work schedules that modifies a biological rhythm; occurs when internal circadian rhythms (entrained to the time zone of departure) are "out of synch" with environmental cues at the time zone of arrival.

kilocalorie (kcal)—The thermal energy required to raise the temperature of 1 L of water 1°C.

lung squeeze—A form of underwater pressure trauma; a diver descends to a depth at which internal lung pressure (i.e., from blood to alveoli) exceeds the pressure within the alveoli; this causes a movement of fluid into the lung and greatly reduces O_2 diffusion from the lung to the blood.

maximal muscular strength—The maximal force (in units of kilograms, pounds) that a muscle can exert during one maximal repetition (1 RM).

mediators—Biological, social, and psychological modifiers that act on **stressors** to alter the level of physiologic strain experienced.

medical climatology—The study of the influences of natural climates on health.

metabolic acclimatization—A type of metabolic adaptation to cold exposure; change(s) in either the type of nutrient used for heat production or the metabolic pathway.

metabolism—The sum of all chemical and physical changes that occur within the human body, including energy transformations.

meteorology—The study of weather and weather forecasting.

middle-ear squeeze—A form of underwater pressure trauma that occurs because the eustachian tube is closed (due to a mucus plug or inflammation); at depths of 1.3-5.3 m, the external water pressure becomes greater than the pressure in the middle ear and rupture of the tympanic membrane is possible.

mucociliary transport—Coordinated, rhythmic movement of microscopic hairs (e.g., cilia) lining the bronchial tubes; moves mucus and foreign bodies out of the lungs and airways.

muscular endurance—The ability to sustain continuous contractions at submaximal intensity for several minutes to hours.

nasal allergy—Hypersensitivity in the nasal cavity to normally harmless, inhaled substances.

neutrophil—A white blood cell involved in immune function; destroys bacteria and cell debris via **phagocytosis**.

nitrogen dioxide (NO_2)—An air pollutant discharged by motor vehicle or aircraft engines and cigarette smoking; causes a mild irritation of the upper respiratory tract and impairment of mucociliary activity in bronchial tubes; individuals who have respiratory disorders are more susceptible to NO_2.

nitrogen narcosis—An increased level of N_2 in blood that causes many exotic physical and mental symptoms similar to alcohol intoxication; this condition generates anesthetic-like euphoria, overconfidence, poor judgment, and a slower reaction time; many divers have died from nitrogen narcosis because of serious errors in diving techniques and accidents.

nonshivering thermogenesis—Metabolic heat generated by all sources except shivering.

norepinephrine (noradrenaline)—The neurotransmitter of most of the sympathetic nervous system; also released by the adrenal medulla; prepares the body for the "fight or flight" response and mobilization/expenditure of energy (i.e., glucose and free fatty acid mobilization in blood); increases blood pressure, cardiac output, cell metabolism, and mental activity.

osmolality—The concentration of a sample; affected by all dissolved particles in a standardized volume of fluid.

oxygen toxicity—Oxygen may become a poison to the central nervous system (CNS), lungs, and eyes when present at high levels for a prolonged period.

ozone (O_3)—In large cities, O_3 is an air pollutant formed by the action of sunlight on automobile exhaust fumes; ozone is one of the main constituents of smog; inhalation of O_3 impairs pulmonary function, causes respiratory discomfort, and increases the number of clinical symptoms; these responses are exacerbated by exercise.

parasympathetic nervous system—A division of the autonomic nervous system; functions to conserve or restore energy.

peak muscular power—The maximal power (in watts) that a muscle can exert.

pernio—A minor form of cold injury that affects the legs, toes, hands, and ears; a severe cold-induced skin vasoconstriction causes lack of oxygen in cells and limb swelling; similar to chilblain.

peroxyacetyl nitrate (PAN)—An air pollutant formed by the action of sunlight on automobile exhaust fumes; one of the common constituents of smog; an eye irritant with a distinctive odor that results in blurred vision and eye fatigue.

phagocytosis—Ingestion and digestion of bacteria and cell debris by specialized cells known as phagocytes.

pneumothorax—A form of underwater pressure trauma in which lung tissue bursts and air is forced through the alveoli; this causes a pocket of air to form in the chest cavity, between the inner chest wall and the outer lung tissue.

polycythemia—Increased production of red blood cells leading to an increased hematocrit.

power—Power is the product of strength and speed of movement; power = (force · distance)/time or (force · velocity); power is the explosive aspect of strength.

primary pollutant—An air pollutant that affects the body directly from its source; includes carbon monoxide, nitrogen oxides, sulfur oxides, and particulates.

pulmonary ventilation (V_E)—The volume of air that passes into and out of the respiratory system per minute; also known as minute volume.

radiation—Transfer of energy waves that are emitted by one object and absorbed by another (e.g., solar energy from sunlight, radiant heat from the ground).

relative humidity (rh)—An index of the water content of air, relative to a 100% saturated air sample (i.e., one that can hold no additional moisture).

RNA—Ribonucleic acid; a molecule that stores hereditary information, serves a structural role in cells, and acts as an **enzyme** in chemical reactions.

secondary pollutant—A pollutant formed in the air by the interaction of **primary pollutants** with ultraviolet light, other compounds, or other primary pollutants; includes aerosols, ozone, peroxyacetyl nitrate.

serotonin—A brain neurochemical that directly affects the pituitary and its secretion of the hormones ACTH and prolactin.

serotonin irritation syndrome—A medical syndrome believed to result from atmospheric effects on the brain, specifically the action of positive air ions; physiological effects include smooth muscle contraction, constriction of blood vessels, and increased respiratory rates.

shivering thermogenesis—Heat generated by muscle during shivering.

shivering threshold—The body temperature at which shivering begins.

sinus squeeze—A diving injury to the membrane of the sinus cavity, including intense pain and bleeding, caused by increased pressure during descent.

specific gravity—The density (mass per volume) of a urine sample, in comparison to pure water. Any fluid that is denser than water has a specific gravity greater than 1.000. Normal urine values usually range from 1.013 to 1.029 in healthy adults. During dehydration, urine specific gravity exceeds 1.030.

strain—Physiologic responses of organs and systems (e.g., increased heart rate, core body temperature, blood pressure, and breathing rate) that express the presence of **stressors**.

stress—The body's recognition of threats to the stability of cells and organs.

stressors—Influences that throw your body out of homeostatic balance (e.g., unpleasant or noxious stimuli).

suit squeeze—Pinching or reddening of skin that becomes trapped beneath a fold of a diver's suit during descent.

sulfur dioxide (SO_2)—An air pollutant; formed when sulfur-containing fossil fuels are burned; exerts its influence mainly as an upper respiratory tract irritant.

sympathetic nervous system—A division of the autonomic nervous system; it prepares the body for the "fight or flight" response and mobilization/expenditure of energy.

synchronizer—An external cue that resets the duration of a biological rhythm; synonymous with the terms **zeitgeber** and **entraining agent**.

trench foot—A form of cold injury caused by prolonged exposure of wet feet to a temperature of 0.5-10°C (33-50°F); nerves and blood vessels are damaged; similar to immersion foot.

vasoconstriction—Blood vessel constriction that decreases blood flow to a specific area of the body; this increases peripheral resistance to blood flow.

V̇O₂max—Maximal oxygen uptake (e.g., maximal aerobic power); the best single measurement of cardiorespiratory endurance and aerobic fitness; measured by analyzing expired air (i.e., oxygen and carbon dioxide) during an exercise test to exhaustion.

zeitgeber—Derived from a German word that means "time giver"; an external cue that resets the duration of a biological rhythm; synonymous with the terms **entraining agent** and **synchronizer**.

SUGGESTED READINGS

Baker, P.T. 1974. An evolutionary perspective on environmental physiology. In *Environmental physiology*, edited by N.B. Slonim. St Louis: Mosby.

Cannon, W.B. 1929. Organization for physiological homeostasis. *Physiological Reviews 9*: 399-431.

Mrosovsky, N. 1990. *Rheostasis: The physiology of change.* New York: Oxford University Press, 3-5.

Sherrington, C.S. 1941. *Man on his nature.* Cambridge, England: University Press, 84.

INDEX

ABOUT THE AUTHOR

Lawrence E. Armstrong, PhD, Fellow of the American College of Sports Medicine (ACSM), is an associate professor in the Department of Kinesiology, Human Performance Laboratory, at the University of Connecticut. He has received the Aerospace Medical Society's Environmental Science Award (1986), and the National Strength and Conditioning Association's (NSCA's) Presidential Award for contributions to the NSCA Journal in Environmental Physiology (1989 and 1994). Since 1982, he has written more than 60 research articles for scientific journals and nearly 50 articles for educational and consumer publications. He also has contributed chapters to a number of books and has coauthored numerous articles for government technical reports.

Armstrong likewise has "hands-on" experience relative to extreme environments. In addition to completing 14 marathons and climbing Mt. Washington three times, he has collected research data in the medical tent for the Boston Marathon and contributed to ACSM position stands on fluid replacement during exercise as well as on heat and cold illnesses contracted during distance running. He graduated cum laude as a scholar-athlete from the University of Toledo in 1971 with a BEd in Biology and Comprehensive Science, earned a MEd from Toledo in 1976, and a PhD from Ball State University in 1983 as a student of David L. Costill. He is a past president of the New England Chapter of the ACSM and conducted numerous research studies as a physiologist at the Research Institute of Environmental Medicine in Natick, MA, from 1983 to 1990. Armstrong lives in Mansfield Center, Connecticut.

*You'll find
other outstanding
...al activity resources at*

www.Hum......netics.com

In the U.S. call

1-800-747-4457

Australia.............................. 08 8372 0999
Canada 1-800-465-7301
Europe...................... +44 (0) 113 255 5665
New Zealand................... 0064 9 448 1207

HUMAN KINETICS
The Information Leader in Physical Activity
P.O. Box 5076 • Champaign, IL 61825-5076 USA